RAISED BED GARDENING

SIMPLIFIED

A Clear, Step-by-Step System

for Growing Vegetables in Raised Garden Beds, No Experience Required

TALI MOSS

CONTENTS

INTRODUCTION:

What You'll Be Growing Soon

(And Why This Book Takes a Different Approach)

If you have ever tried to grow vegetables before and ended up with a half-empty bed, a handful of struggling plants, and the uncomfortable sense that everyone else seems to find this easier than you do, it is worth saying this upfront. Most people do not fail at gardening because they are careless or incapable. They fail because the advice they are given assumes far more time, interest, and tolerance for complexity than real life allows.

This book is not written for people who want a new hobby, a weekend obsession, or a reason to spend hours comparing soil blends and planting charts. It is written for people who want to grow food, ideally without turning their spare time into a research project or their backyard into a source of ongoing stress. People who are busy, practical, and quietly skeptical of anything that promises easy results while demanding a lot of effort behind the scenes.

Most raised bed gardening books miss the mark in predictable ways. Some overwhelm you almost immediately, piling on options, terminology, and contradictory rules until even simple decisions start to feel risky, and the raised bed you were excited about becomes something you keep meaning to get to once you have figured it out properly. Others swing too far in the opposite direction, offering cheerful reassurance without explaining the few decisions that actually matter, so when things do not grow the way you expected, you are left unsure what went wrong or how to fix it.

If you have tried before, followed the instructions, and still watched things fail, this book assumes you did not do anything wrong.

A quick example, so you know what I mean. One beginner I spoke to had tried container gardening twice, watched everything struggle, and decided they just "weren't a plant person." They set up one small raised bed using the simple mix in this book, planted fast greens and a few herbs, and saw obvious growth within two weeks. Not perfection, not some fairy tale harvest overnight, but enough progress to prove the system was working and to stop the endless second guessing.

This book was built to avoid both of those traps by following one guiding idea that shows up again and again, whether the topic is soil, planting, watering, or troubleshooting. This book runs on one rule that shows up again and again: the **No Overthinking Rule**. Do not overthink what does not deserve it. You will see that rule applied to soil, planting, watering, feeding, and troubleshooting, because beginners do not need more information, they need fewer decisions.

Every recommendation you will see here has been filtered through a simple question. What will reliably work for most beginners, in most situations, without requiring constant attention or a high tolerance for trial and error. If something adds complexity without meaningfully improving your chances of success, it does not make the cut. If there is a good enough option that works just as well as the more complicated one, that is the option you will be pointed toward.

This is not about cutting corners or pretending plants do not have basic needs. It is about focusing your effort where it actually pays off and letting go of the rest without guilt.

This book will walk you through choosing vegetables that are forgiving rather than temperamental, setting up a raised bed in a way that makes everything easier later on, planting without memorizing charts or cross-referencing calendars, and keeping plants alive with routines that fit into real schedules instead of ideal ones. When something does not look right, and at some point something will not, you will have a clear and fast way to figure out what is happening and what to do about it, without spiraling into panic or assuming you have ruined everything.

What this book will not do is try to turn you into a gardener by identity, push expensive tools or products you do not need, or imply that enthusiasm and perfection are required for success. It also will not pretend that nothing ever goes wrong. Plants die sometimes. Weather does what it wants. That is part of the deal, and it is not a personal failure.

If you follow the approach in this book, most beginners can expect to see real and visible progress within the first month. Fast greens that can be harvested more than once, herbs that earn their space by actually getting used, and vegetables that are clearly growing and heading in the right direction. Not a flawless garden, and not something you would photograph for a magazine, but enough success to know that the system is working and that you do not need to constantly intervene to keep it that way.

Raised beds are not magic, but they are forgiving when they are set up simply. They drain better, warm up faster, and give you far more control than planting straight into the ground, which is why they are such a good option for beginners who want results without having to fight their soil or start from scratch every season. They give you room to make small mistakes without everything collapsing, which is often the difference between sticking with gardening and giving up on it altogether.

Most gardening advice assumes you will reshape your life around your garden. This book works the other way around, fitting gardening into a life that already includes work, family, limited energy, and better things to do than constantly babysit plants. This is written for people who want vegetables, not another thing to manage. You do not need to love the process for it to work, and you do not need to look anything up beyond these pages. You just need a clear starting point, fewer decisions, and permission to keep things simple for as long as you want.

If decision fatigue has ever stopped you from starting, the structure of this book is designed to remove that friction. By the time you finish the next chapter, you will not be wondering what you should grow or whether you are setting yourself up to fail. You will know exactly where to start, what to skip for now, and why that restraint is one of the most practical choices you can make.

PART I:

Decide What You're Growing

(So Everything Else Is Easier)

\

Chapter 1:

Before You Choose Anything: How This Actually Works

Most beginners do not fail because they choose the wrong vegetables. They fail because they choose **too many, too early**, and pick crops that punish inexperience instead of forgiving it.

Before you choose a single plant, before you buy seeds or seedlings, before you measure sun or open a shopping tab, there is something important to clear away first, and that is the quiet pressure many people carry into gardening without realizing it, the sense that they need to get everything right from the start or the whole thing will be a waste of time, money, and hope.

If you have ever felt overwhelmed before you even began, that feeling did not come from laziness or lack of ability. It came from the way gardening advice is usually presented, as a long list of rules, optimizations, and ideal conditions that make it sound like success belongs to people with more time, more space, or more confidence than you have right now. This chapter exists to undo that pressure, not by giving you more information, but by changing how you think about what actually matters in the beginning, so that every decision that follows becomes easier instead of heavier.

Raised bed gardening works best when it is treated as a system designed to support learning, not a test you pass or fail. The mistake most beginners make is assuming that success depends on precision, when in reality it depends on repeatability, on choosing an approach you can maintain without constant correction or emotional effort. This book is built around that idea, and before we go any further, it helps to understand how that changes everything.

Why Gardening Feels Hard Before You Even Start

For many people, the hardest part of gardening is not planting or watering. It is standing at the beginning, surrounded by advice, and trying to figure out which version of gardening they are supposed to follow. One source says to start seeds indoors weeks in advance. Another insists seedlings are the only sensible option. Someone else swears that unless your soil mix is perfect, nothing will grow. Add in climate charts, spacing diagrams, fertilizer schedules, and warnings about what not to do, and it starts to feel like a system designed to expose mistakes rather than prevent them.

That feeling of being behind before you have even started is what causes most beginner gardens to fail, not because people do not care, but because the mental load becomes too heavy to carry alongside normal life. When gardening feels like homework, it gets postponed. When it feels like a performance, it gets abandoned the moment something goes wrong.

Raised beds reduce the physical complexity of gardening, but they also need a mental reset to work as intended. The real advantage of a raised bed is not just better drainage or nicer soil. It is that you are working within clear boundaries, with fewer variables, which means you can stop trying to solve everything at once. But that advantage only shows up if you give yourself permission to approach gardening as a process rather than a performance.

What This Book Means by "It Works"

When people say a garden works, they often picture abundance, perfect rows, and baskets of produce that look good enough to photograph. That image is appealing, but it is not a useful benchmark for beginners, because it sets the bar so high that anything less feels like failure.

In this book, working means something much simpler and far more achievable. It means plants grow without constant intervention. It means you can miss a day or two without everything collapsing. It means you understand what you are seeing well enough that you do not panic when something looks imperfect. It means the garden fits into your life instead of competing with it.

If you grow some food, even modestly, and the process feels manageable, the system is working. If you learn enough in the first season that the second feels calmer, the system is working. If you feel more confident at the end than you did at the beginning, even if everything did not grow perfectly, the system is working.

This matters because once you change the definition of success, you stop chasing outcomes that are not necessary for satisfaction or progress, and you start noticing the small, real wins that actually build confidence.

The Difference Between Productive Gardening and Perfect Gardening

Perfect gardening is optimized for appearance and control. It focuses on exact spacing, ideal timing, and constant adjustment. It assumes a high level of attention and rewards people who enjoy fine-tuning details.

Productive gardening is optimized for results with minimal friction. It focuses on light, water, and time. It accepts unevenness and variation. It rewards consistency, not precision.

Most beginner advice is written from the perspective of perfect gardening, even when it claims to be beginner friendly. It assumes the reader wants to optimize, tweak, and improve continuously. For some people, that is enjoyable. For many others, it is exhausting.

This book is firmly in the productive gardening camp. That does not mean careless or uninformed. It means choosing approaches that tolerate mistakes, simplify decisions, and allow learning to happen without penalty. It means doing fewer things, but doing them well enough that plants can take over most of the work themselves.

Once you understand this difference, it becomes much easier to let go of advice that sounds impressive but does not actually serve your goals.

What Actually Matters in the First Season

In the first season, there are only a few things that truly determine whether your raised bed will succeed.

Plants need enough light to grow. They need soil that drains but holds moisture. They need water applied with some consistency. They need time to adjust and develop.

That is the list.

Everything else, from exact spacing to ideal fertilizer schedules, has far less impact than it is often given credit for, especially early on. This is not an argument for ignoring basic guidance. It is an argument for prioritizing the few factors that move the needle, instead of spreading your attention across dozens of details that do not.

Beginners often assume that missing a specific step or choosing the wrong variety will ruin everything. In reality, plants are far more adaptable than advice makes them seem. They respond to patterns more than perfection. If light, water, and soil conditions are reasonably stable, plants will usually find a way to grow, even if other elements are less than ideal.

Understanding this reduces fear because it shifts your focus from avoiding mistakes to creating a supportive baseline, and that baseline is easier to maintain than a constantly optimized setup.

Why Starting Smaller Leads to Faster Confidence

One of the most counterintuitive truths in gardening is that doing less at the beginning often leads to better outcomes, not because the plants need less, but because you do.

When people start with too many plants or too large a setup, they create a situation where every small issue feels multiplied. Watering becomes more complicated. Problems feel harder to diagnose. Missed days feel more consequential. The garden stops feeling approachable.

Starting smaller changes the emotional equation. You can observe without rushing. You can notice changes without feeling overwhelmed. You can make small adjustments and actually see their effects. This feedback loop builds confidence quickly, which is what keeps people going long enough to succeed.

This book will repeatedly steer you toward choices that limit scope early on, not because you are incapable of more, but because confidence compounds faster than complexity. Once you feel grounded, scaling up becomes a choice rather than a challenge.

How This Book Reduces Decisions for You

One reason gardening advice feels heavy is that it asks you to make too many decisions too early. Which plants. Which soil mix. Which bed size. Which fertilizer. Which watering schedule. Each choice carries implied risk, and risk accumulates.

Where possible, this book removes decisions instead of presenting options. It offers default paths that work for most people most of the time. When alternatives are presented, they are framed as optional, not required, and the reasons for choosing one over another are explained simply.

This matters because decision fatigue is real. When you reduce the number of choices you have to make, you free up energy to observe, learn, and enjoy the process. You also reduce the urge to second-guess yourself later, because you are following a system designed to be forgiving, not a custom plan that depends on perfect execution.

If you ever feel unsure, the safest move in this book is usually the simplest one. That is not an accident. It is how the system is designed.

What "Good Enough" Looks Like in Real Gardens

Good enough is a bed that gets sun for most of the day, even if not all day. It is soil that drains when watered, even if it does not look perfectly uniform. It is watering when the soil needs it, even if that schedule shifts with weather and life. It is plants that grow at slightly different rates, some thriving, some merely surviving.

Good enough is not sloppy. It is functional. It leaves room for learning without punishment.

Many people never experience this version of gardening because they aim for ideal from the start and feel discouraged when reality does not match it. By aiming for good enough, you create conditions where progress feels visible and sustainable.

This is especially important in raised beds, where the contained environment amplifies both successes and mistakes. A forgiving approach allows the system to settle into balance instead of swinging between extremes.

A beginner-friendly choice is not the same as a popular or impressive one, and confusing the two creates unnecessary frustration.

Letting Go of the Fear of Wasting Time or Money

A common, unspoken fear for beginners is the fear of waste. What if the plants die. What if the soil was wrong. What if the bed sits empty. That fear makes people hesitant to start or quick to abandon the project at the first sign of trouble.

It helps to reframe the first season not as a test, but as an investment in understanding. Even if some plants fail, the soil improves. The bed remains. Your knowledge increases. Very little is truly lost.

Raised beds are particularly good at this because they are reusable systems. Soil can be amended. Layouts can change. Plants can be replaced. Each attempt builds on the last rather than starting from zero.

When you remove the idea that everything has to work perfectly to be worth doing, you make it much easier to keep going long enough to succeed.

What This Book Will Not Ask You to Do

This book will not ask you to track data obsessively, follow rigid schedules, or chase optimal results. It will not require specialized tools, constant feeding, or daily intervention. It will not assume unlimited time, physical ability, or enthusiasm.

Instead, it will ask you to notice a few key signals, respond calmly, and allow the system to do most of the work. It will encourage you to skip steps that do not add value and to trust processes that unfold slowly.

This is not about lowering standards. It is about aligning effort with outcomes so that gardening remains something you return to willingly rather than something you avoid.

How to Move Forward Without Pressure

As you move into the next chapter, you will be asked to decide what to grow first. That decision will feel lighter if you carry one simple idea with you: the goal is not to get it right, but to get it going.

The choices you make now are starting points, not commitments. You can adjust, replace, and refine as you learn. Plants are not permanent. Beds are flexible. Knowledge accumulates.

If you ever feel stuck, choose the option that feels easiest to maintain rather than the one that promises the best result. Ease is not a compromise. It is a strategy.

What to Do Next

Choosing what to skip early is one of the most practical decisions a beginner can make. Before choosing plants, pause and check in with yourself. Notice whether you feel rushed or pressured, and remind yourself that nothing here is irreversible. The system you are building is designed to support learning, not demand perfection.

In the next chapter, we will narrow your choices in a practical way, focusing on plants that build confidence rather than frustration, so you can take your first steps with clarity and calm, knowing that you are not behind, and you do not need to catch up to anyone else to succeed.

No Overthinking Rule #1

If you catch yourself researching instead of doing, **stop** and do the **simplest next step**.

Chapter 2:

What to Grow First

(And What to Skip for Now)

If you want the shortest path to success, you do not need variety. You need a small group of vegetables that grow quickly, tolerate mistakes, and give clear feedback when something is wrong. Everything in this chapter is filtered through that goal. If you want to remove guesswork, treat what follows as a set of decisions already made for you.

Many people feel excited about the idea of growing their own vegetables and, at the same time, quietly intimidated by the possibility of getting it wrong. If you are one of them, you are exactly the kind of reader this chapter is written for, because the very first decisions you make in a raised bed garden have far more influence on how the entire experience feels than most people realize, not because they determine whether plants live or die, but because they determine whether *you* feel capable, calm, and willing to keep going.

Most beginner gardening advice treats plant choice as a technical problem, as though success depends on matching the right crop to the right season, the right soil, and the right spacing, when in reality the bigger issue is emotional, especially at the start, because certain plants quietly reward uncertainty with visible progress, while others amplify every small hesitation until the garden starts to feel like a test you are failing in slow motion.

This chapter is not about choosing the most impressive vegetables, the ones people post photos of or brag about, it is about choosing plants that cooperate with a beginner's reality, plants that grow in a way that builds confidence rather than drains it, plants that allow you to learn by doing without punishing you for not knowing everything yet.

If You've Failed Before, Read This First

A large number of people who come to raised bed gardening are not trying something new for the first time, they are trying again, often after one or more experiences that quietly convinced them they were bad at gardening, disorganized, impatient, or simply not cut out for it, even if they never said those words out loud.

If that's you, it's important to say this clearly before going any further: most gardening "failures" are not failures of effort or intelligence, they are failures of sequencing, where someone starts with plants that demand precision, patience, or consistency before they have had a chance to build any confidence or intuition at all.

Planting slow-growing, high-maintenance vegetables as a first attempt creates a distorted narrative, because weeks can pass with little visible change, problems feel mysterious rather than solvable, and every piece of advice seems to contradict the last, until eventually the garden becomes something you avoid rather than engage with, not because you don't care, but because caring starts to feel uncomfortable.

Raised beds do not magically fix this on their own, but they do give you a chance to reset the story, because when you pair them with forgiving plant choices, you create conditions where progress shows up early, mistakes are survivable, and learning happens gradually rather than all at once.

This book assumes you are not starting from zero, even if it feels that way, it assumes you are starting again, this time with better sequencing, lower pressure, and permission to choose plants that want to succeed with you rather than in spite of you.

Why Plant Choice Matters More Than Skill

Many beginners believe that gardening success comes from mastering technique, watering schedules, feeding routines, and seasonal timing, and while those things do matter eventually, they are not what determine whether a first season feels successful or discouraging.

The real determinant is how much feedback the garden gives you early on.

Plants that grow quickly, respond visibly to care, and tolerate small inconsistencies create a feedback loop where effort feels worthwhile, because you can see that something is happening, even if you are not sure exactly why yet. Plants that grow slowly, hide their progress underground, or require precise conditions create the opposite loop, where effort feels disconnected from outcome, and uncertainty multiplies rather than resolves.

Skill develops through repetition and observation, not through memorization, and repetition only happens when the garden feels approachable enough to return to regularly. If a plant does not give visible feedback within a few weeks, it is not a good first choice.

Choosing the right plants at the beginning is not about playing it safe, it is about choosing a learning curve that slopes gently rather than steeply.

What Makes a Plant Beginner-Friendly

Instead of thinking in terms of "easy" and "hard," it's more useful to think in terms of forgiveness.

Beginner-friendly plants tend to share a few quiet characteristics: they germinate or establish relatively quickly, they tolerate a range of temperatures, they do not demand exact spacing, and they recover well from small mistakes in watering or handling. They also tend to offer harvests in stages, rather than all at once, which removes the pressure of getting timing exactly right.

These plants allow you to make decisions, observe the results, and adjust, without forcing you to start over every time something doesn't go as planned.

Plants that only reward perfection, even if they are technically simple, are better left for later, when confidence and familiarity are already in place. For your first raised bed, treat five crops as the upper limit. Fewer is often better.

Leafy Greens: Why They Build Confidence So Fast

Leafy greens are often recommended to beginners, but the reason they work so well is rarely explained clearly.

Lettuce, spinach, arugula, and similar greens grow quickly enough that you don't have time to convince yourself you've failed before something happens, they tolerate cooler temperatures that would stall other crops, and they are far less sensitive to spacing errors than most people expect, especially in raised beds where soil quality and drainage are already improved.

They also introduce harvesting gently, because you don't have to wait for a single final moment, you can remove a few leaves at a time, see that the plant survives, and slowly build trust in your own judgment. If you are unsure what to plant first, start with leafy greens.

Perhaps most importantly, leafy greens make the bed look alive early, which sounds superficial, but is actually critical, because visual progress does more to reduce anxiety than almost any piece of advice ever will.

Herbs That Are Worth Growing Early

Herbs occupy a unique place in beginner gardens, because they deliver usefulness in very small quantities, which means even modest success feels meaningful.

Basil, parsley, chives, cilantro, and similar herbs establish quickly, tolerate partial harvesting, and often respond to being cut by growing back more vigorously, which reinforces the idea that interacting with the garden is not inherently risky.

They also tend to fit easily into meals, even if you cook infrequently, which strengthens the connection between the garden and daily life, rather than making it feel like a separate project that demands special attention.

Not every herb is equally forgiving, but starting with a few reliable ones creates early wins without increasing complexity. Limit yourself to two or three herbs at most in the first bed.

Vegetables That Pay You Back Quickly

Some vegetables are especially helpful early on because they shorten the gap between effort and reward.

Radishes, certain Asian greens, and green onions are good examples, because they show visible progress quickly and mature fast enough that you don't have to hold uncertainty for long periods. Even if the harvest is small, the experience of planting something and using it within a few weeks does more to build confidence than waiting months for a single outcome.

Fast results do not make you impatient, they make you willing to keep going. When choosing between two vegetables, pick the one that matures faster.

Taste, Yield, and Effort: You Can't Optimize All Three

One of the quiet sources of frustration for beginners is trying to optimize everything at once, without realizing that most plant choices involve trade-offs.

Taste, yield, and effort exist in a kind of triangle, and while it is possible to optimize two at a time, trying to optimize all three usually leads to disappointment. Some vegetables taste incredible but demand attention and timing. Others produce large yields but require feeding, space, and patience. Still others are easy to grow but offer more modest harvests.

Early on, effort should be your primary filter. If a vegetable requires regular feeding, precise timing, or constant monitoring, it is not a priority in the first season.

Choosing plants that require less attention, even if they are not the most exciting or productive on paper, creates space to learn, and learning is what eventually allows you to pursue taste or yield more intentionally later.

There is nothing wrong with choosing "boring but easy" foods at the start, especially when the alternative is choosing impressive plants that make the garden feel like work.

The Quiet Problem With "Beginner Favorites"

Tomatoes, peppers, and cucumbers are often labeled as beginner plants, and they can be grown successfully, but they come with expectations that are not always stated clearly.

They usually require consistent warmth, regular feeding, and some form of support, and when conditions are right, they are extremely rewarding. When conditions are not right, especially early in the season, they can dominate your attention and make you feel as though the entire bed is failing, even if other plants are doing fine.

This does not mean you should avoid them entirely, it means you should be intentional.

If you choose to grow them early, pair them with easier plants so your confidence does not hinge on one demanding crop. Do not let any single demanding plant dominate the bed.

What Looks Easy but Often Isn't

Some vegetables seem simple because they are familiar, but familiarity does not always equal forgiveness.

Carrots require loose, well-prepared soil and patience during germination. Onions take a long time and offer little visual feedback early on. Corn requires space and coordination that raised beds do not always provide efficiently.

These plants are not impossible, they are simply less cooperative when you are still learning how your bed behaves, how moisture moves through the soil, and how much attention you realistically want to give.

Skipping them early is not avoiding challenge, it is sequencing it. These are not bad vegetables, they are simply better added once you have one successful season behind you.

Emotional Wins Versus Practical Wins

Not all success in gardening is measured in pounds of produce.

An emotional win is a plant that grows visibly, responds well to care, and reassures you that you are capable of learning this. A practical win is a plant that produces a lot of food.

Early on, emotional wins matter more, because confidence reduces over-intervention, encourages consistency, and keeps the garden from becoming something you avoid when life gets busy.

Practical wins become easier once you are no longer gardening from a place of doubt.

No Overthinking Rule #2

If You Only Grew Five Things

You only need to make five decisions to start a raised bed. Anything beyond that is refinement, not readiness.

If you want to skip decision making entirely, copy this structure and move on. If your goal were to make the first season as manageable as possible, growing a small number of reliable plants would be enough.

A combination of leafy greens, a couple of herbs, and one or two faster vegetables gives you variety without complexity, and allows you to practice planting, watering, harvesting, and observation without juggling too many variables at once.

If you cook often, choose plants you already use regularly, even if they are not exciting. If you cook rarely, choose plants that are forgiving and flexible, rather than ones that demand specific recipes. If visual reassurance matters to you, prioritize plants that fill space and show growth early.

The plants themselves matter less than how they make the garden feel.

Seeds or Seedlings: Choosing Based on Comfort

At this stage, there is no need to take a philosophical stance on seeds versus seedlings.

Seeds are cheaper and offer more variety, but they require patience and trust during the early stages when nothing is visible. Seedlings provide immediate feedback and shorten the waiting period, which can feel calming if uncertainty makes you hesitate.

Many successful gardens use both.

If you feel stuck deciding, choose the option that reduces stress. Comfort is a valid criterion at this stage. Comfort is a valid criterion.

Why Starting Small Is Not Playing It Safe

Starting small is often misinterpreted as a lack of ambition, when in reality it is a strategy for learning faster.

Every additional plant type adds decisions, variables, and opportunities for confusion. Fewer plant types make patterns easier to see and outcomes easier to understand.

A small, successful bed teaches you more than a large, overwhelming one, because it allows you to notice what works without constant distraction.

Ambition is not about doing everything at once, it is about sequencing challenges in a way that builds momentum rather than drains it.

A Common Beginner Scenario, Revisited

Imagine planting an entire bed with slow-growing, high-maintenance plants and waiting weeks with little visible change, adjusting constantly, and slowly losing confidence.

Now imagine that same bed planted with fast greens, forgiving herbs, and one or two longer-term crops, where you are harvesting something within weeks, the bed looks alive, and the slower plants no longer feel like a gamble because they are not carrying the emotional weight of the entire season.

Nothing about your effort changed. Only the plant choices did.

What to Do Next

Before moving on, take a moment to decide what kind of experience you want from your first season.

If you want reassurance, choose plants that grow quickly and forgive mistakes. If you want challenge, add one or two demanding plants, but don't let them dominate the bed. The default choice in this book is always the simplest option, unless there is a clear reason not to use it.

Write it down or note it mentally, not as a commitment, but as a starting point.

In the next chapter, we'll talk about where to place your raised bed in a way that supports these choices without obsessing over perfection.

For now, remember this.

You don't need to prove anything with what you grow, you only need to choose plants that make it easier to learn, easier to return, and easier to trust yourself as you go.

PART II:

Set It Up Once

(So It's Easy Later)

Chapter 3:

Where to Put Your Raised Bed

(Without Obsessing)

One of the fastest ways to stall a gardening project is to spend too long deciding where the raised bed should go. It feels like a serious decision, so people research it like one. They read about sun angles, seasonal light changes, ideal orientations, and all the ways you can supposedly get it wrong.

Then nothing gets built.

This chapter exists to pull you out of that loop.

Where you place your raised bed matters, but not in the way the internet often makes it seem. You are not choosing a forever location that will determine whether your vegetables thrive or fail. You are choosing a workable spot that lets plants get enough light, lets you reach them easily, and fits into your actual life.

That is enough.

If you remember nothing else from this chapter, remember this. A raised bed in a decent spot that you actually use will outperform a theoretically perfect bed that makes you hesitate every time you look at it.

The Truth About Sunlight, Without the Drama

Most vegetables prefer full sun. You will see this phrase everywhere, usually followed by a number like six to eight hours a day. That information is not wrong, but it is often presented in a way that makes beginners think anything less is a deal-breaker.

It is not.

Full sun simply means that the plants receive direct sunlight for a good portion of the day. It does not mean the light has to be uninterrupted, perfectly angled, or measured with an app. It does not mean your garden fails if a fence casts a shadow in the afternoon or a tree blocks the early morning light.

In most US climates, vegetables do best when they get solid light during the middle of the day. Morning sun is especially valuable because it dries leaves and gets plants going early. Afternoon sun is helpful too, but it is also when heat stress can show up in warmer areas.

If your raised bed gets somewhere between five and seven hours of decent sun, you are already in a workable range. If it gets more, great. If it gets slightly less, you can still grow plenty of things, especially the beginner-friendly vegetables you chose in Chapter 1.

The mistake is thinking you need perfect sun to begin. You do not.

Why "The Best Spot" Is Often the Wrong Goal

It is tempting to hunt for the best possible location. The brightest, most open, most textbook-perfect part of the yard. On paper, that sounds logical.

In real life, that spot is often inconvenient.

It might be far from the house. It might require walking across uneven ground. It might be in a place you do not naturally look at or pass by during the day. Over time, those small inconveniences add friction. You water less consistently. You notice problems later. You stop checking the bed as often.

A slightly less sunny spot that you see every day usually wins.

Gardening success is tied closely to attention, not effort. You do not need to work harder. You need to notice sooner. Beds that are easy to see and reach get more of that quiet, casual attention that keeps things on track.

So instead of asking where the sun is best, it helps to ask a different question. Where is the bed most likely to become part of my routine.

The "Good Enough" Placement Rule

Here is a rule that removes most of the stress from this decision.

If a spot meets these three conditions, it is good enough to start.

First, it gets several hours of direct sun on most days. Not perfectly measured. Just enough that you can tell plants are receiving light.

Second, you can access it easily. You do not have to squeeze past obstacles or make a special trip just to check on it.

Third, it fits the space you actually have. Not the space you wish you had.

That is it.

If a location meets those criteria, you are allowed to stop looking for something better.

Perfection is not required for vegetables to grow. Consistency is far more important.

Working With Small Yards, Courtyards, and Tight Spaces

Not everyone has a wide-open yard. Many people are working with courtyards, patios, narrow side yards, or awkward corners. This does not disqualify you from raised bed gardening.

Raised beds are flexible by design. They do not need to be placed in the middle of a lawn. They can sit against fences, along walls, or in spaces that would be difficult to garden in the ground.

If your space is small, light becomes more variable. You might have strong sun for part of the day and shade for the rest. This is normal.

In these situations, it helps to remember the plant choices you made earlier. Leafy greens and herbs tolerate partial sun well. They do not demand the same intensity as fruiting crops like tomatoes or peppers.

A bed that gets morning sun and afternoon shade, or vice versa, can still be productive if you choose plants that match the conditions.

Again, this is not about squeezing perfection out of a challenging space. It is about matching expectations to reality.

What to Do If You Only Have One Possible Location

Sometimes the decision is already made for you.

There is one place the bed can go, and that is it.

If that is your situation, you can let go of comparison entirely. The only thing that matters is how you work with that spot.

Here is the simple approach.

Place the bed where it fits. Observe how the light moves over a day or two. Then choose plants accordingly. If the spot is sunnier, you have more options. If it is partially shaded, lean into leafy greens and herbs.

You do not need to force the location to behave like something it is not.

Vegetables are adaptable. They respond to patterns, not ideals.

Dealing With Shade Without Panicking

Shade often sounds scarier than it is.

Dappled shade, partial shade, or afternoon shade are all workable conditions for many vegetables. What matters is whether plants receive some direct light during the day, especially during the growing season.

Heavy, full-day shade is more limiting, but even then, it does not mean nothing will grow. It simply narrows your options.

If your raised bed is near trees, fences, or buildings, watch how the shadows move. Shade that shifts throughout the day is very different from shade that sits in one place constantly.

If the bed receives direct sun at some point during the day, you can usually make it work. You may not grow everything. You will still grow something.

That is enough to begin.

Orientation and Direction, Explained Simply

You may come across advice about orienting raised beds north to south or east to west. This advice comes from larger-scale gardening and farming, where rows of plants and consistent sun exposure matter more.

For a single raised bed, or even a few, orientation is a minor factor.

What matters more is that taller plants do not block shorter ones and that the bed receives light when the sun is strongest in your climate. You can address the first issue later when planting. The second is usually solved simply by choosing a reasonably open spot.

If your bed ends up facing a different direction than recommended in a guide, it is not a problem. Vegetables do not know which way is north.

Ground Surface, Drainage, and What Really Matters

Raised beds are forgiving when it comes to ground conditions, but placement still affects drainage and stability.

If possible, place your bed on relatively level ground. This does not mean perfectly flat. It means stable enough that water does not pool on one side and the bed does not feel like it is shifting.

Avoid placing beds directly in low spots where water collects after rain. Raised beds drain better than ground soil, but constant pooling underneath can still cause issues over time.

If your only option is a hard surface like concrete or pavers, that is workable too. Drainage holes and airflow become more important in those cases, and soil moisture may behave slightly differently, but raised beds can function well in these settings.

The key is not to overcorrect. Minor imperfections are rarely the cause of gardening failure.

Convenience Is a Gardening Skill

This is one of those truths that experienced gardeners learn quietly over time.

The easier your bed is to reach, the better it will perform.

Beds that are close to a water source get watered more consistently. Beds that are near the house get checked more often. Beds that are in your line of sight remind you to harvest, notice changes, and respond early to problems.

None of this requires effort. It requires proximity.

If you are choosing between a sunnier spot far away and a slightly less sunny spot near where you already walk, the closer option often wins in practice.

Gardening thrives on small, regular interactions, not big, occasional efforts.

What Not to Worry About Right Now

It is easy to feel like you should account for every variable before placing your bed. Resist that urge.

You do not need to worry about microclimates yet. You do not need to predict seasonal sun angles. You do not need to factor in future landscaping plans unless they are imminent.

You are allowed to place the bed where it works now.

Raised beds can be moved later if needed. Soil can be adjusted. Plants can be changed.

This is not a permanent commitment.

A Simple Way to Decide and Move On

If you are still torn between two or three spots, here is a calm way to decide.

Stand in each location at a time of day when you are usually home. Notice which one feels easiest to access. Notice which one you naturally look at. Notice which one you would not mind walking to with a watering can.

Choose that spot.

This is not a scientific method. It is a practical one. It works because it aligns the garden with your habits instead of asking you to change them.

What to Do Next

Once you have chosen a location, stop revisiting the decision. Mark the spot, mentally or physically, and move on.

In the next chapter, we will look at choosing a raised bed itself. Whether you buy one or build your own, the goal is the same. Pick something sturdy, appropriately sized, and easy to live with.

You do not need the best bed on the market. You need one that lets you get started without second-guessing.

That is where momentum begins.

Chapter 4:

Choosing a Raised Bed You Won't Regret

By the time people reach this point, many of them have already looked at raised beds online. That usually makes things worse before it makes them better.

There are sleek metal beds, chunky wooden ones, modular kits, fabric options, tall beds, low beds, beds with legs, beds with liners, beds with promises of miracle growth. Some are marketed as beginner-friendly. Some are marketed as professional grade. Many are marketed as if the bed itself will do the hard work for you.

It is a lot to take in.

The good news is that raised beds are one of those gardening tools where the basics matter far more than the details. You do not need to find the perfect bed. You need to choose one that is solid, appropriately sized, and easy to live with.

This chapter will help you do exactly that, without turning it into a research project.

Buy or Build: A Calm Way to Decide

Before we talk about materials or sizes, it helps to settle one question early. Are you buying a raised bed, or are you building one.

There is no correct answer here. Both options can work very well. The mistake is thinking that one choice makes you a more serious gardener than the other.

Buying a raised bed makes sense if you want to get started quickly, if tools and DIY projects are not appealing, or if you simply prefer a ready-made solution. A good store-bought bed can last for years and requires very little thought beyond assembly.

Building a raised bed makes sense if you want to save some money, customize the size, or avoid some of the flimsier kits on the market. It also makes sense if you enjoy simple projects and want full control over the materials.

What does not make sense is forcing yourself into a DIY project you will resent, or buying an expensive bed because it looks impressive.

If you feel neutral about both options, choose the one that feels easier to complete in the next two weeks. Momentum matters more than ideals.

We will walk through how to build your own bed in the next chapter. For now, this chapter applies whether you are buying or building.

What a Good Raised Bed Actually Needs to Do

Strip away the marketing, and a raised bed has a very simple job. It needs to hold soil in place, drain well, and stay intact through weather and use.

Everything else is secondary.

A good raised bed should feel stable when you press against it. It should not bow outward once filled with soil. It should not wobble or feel flimsy when you lean in to plant or harvest. These things affect daily use far more than appearance.

It should also be tall enough to give roots room to grow, but not so tall that filling it becomes expensive or unnecessary.

If a bed meets those basic criteria, it will grow vegetables just as well as a much more expensive or elaborate option.

Size Matters More Than Most People Expect

When people regret their raised bed choice, size is often the reason.

Beds that are too small feel cramped quickly. Beds that are too large become hard to manage, especially for beginners who are still learning how much to plant and how to care for it.

There is a reason certain dimensions show up again and again in practical gardens. They work.

For most beginners, a bed that is about four feet wide is ideal. This width allows you to reach the center from either side without stepping into the bed. Stepping on soil compacts it over time, which is something raised beds are meant to avoid.

Length is more flexible. A bed can be six feet long, eight feet long, or longer if space allows. If you are unsure, shorter is often better to start. It is easier to add another bed later than to shrink one that feels overwhelming.

Depth matters too. Beds that are around ten to twelve inches deep work well for most vegetables, especially the leafy greens and herbs you are starting with. Deeper beds are not wrong, but they require more soil, more cost, and more effort to fill. That extra depth does not always translate to better results, especially in the first season.

If you remember one thing about size, remember this. Choose a bed you can comfortably reach into and maintain without stretching, climbing, or stepping inside.

Wood, Metal, Fabric, and Other Options

Raised beds come in several common materials, each with strengths and trade-offs. None of them are inherently bad. Some are simply better suited to beginners than others.

Wood is the most common choice. It is familiar, easy to work with, and blends naturally into most outdoor spaces. When chosen carefully, it can last for years. Wood beds are also forgiving to build and repair, which makes them appealing if you like having control.

Metal beds are increasingly popular. They often come in modular kits and have a clean, modern look. Good-quality metal beds are durable and resistant to rot. They can heat up more in full sun, but in most climates this is manageable, especially if you choose appropriate plants and watering habits.

Fabric beds, sometimes called grow bags, are lightweight and inexpensive. They drain very well and are easy to move. However, they tend to dry out faster and may not last as long. For some people, they are a good temporary solution. For others, they become another thing to replace.

Plastic and composite beds vary widely in quality. Some are sturdy and long-lasting. Others feel flimsy once filled. If you consider this option, stability is the key thing to assess.

Instead of asking which material is best, ask which one fits your space, budget, and tolerance for maintenance. A slightly imperfect bed that you actually use will outperform a theoretically ideal one that feels awkward or annoying.

A Word on Pressure-Treated Wood

This topic causes more confusion than it needs to.

Modern pressure-treated wood sold for residential use in the United States is treated differently than it was decades ago. The chemicals that raised the most concern are no longer used in the same way. Current treatments are generally considered safe for garden use when the wood is rated for ground contact.

Many gardeners still prefer untreated wood or naturally rot-resistant options like cedar. That is a valid preference, especially if it fits your budget.

The practical takeaway is simple. If pressure-treated wood is what allows you to build a sturdy, affordable raised bed, you can use it without panic. If you prefer to avoid it, there are alternatives. Either choice can work.

What matters far more than the wood type is the quality of the soil you put inside and how you care for the plants.

Raised Bed Kits: What to Look For

If you are buying a pre-made raised bed, a few details are worth paying attention to.

First, check the thickness of the material. Thin boards or panels may look fine when empty but bow once filled with soil. This is especially common in longer beds without internal supports.

Second, look at how the corners are constructed. Strong corner connections keep the bed square and stable over time. Weak corners are often the first point of failure.

Third, consider how the bed sits on the ground. Beds that make good contact with the soil beneath tend to be more stable and drain better. Beds that sit on narrow legs or frames can feel less secure, especially when full.

You do not need a premium brand to get a functional bed. You do need something that feels solid once assembled.

Cost Transparency: What Raised Beds Really Cost

It helps to be realistic about cost upfront, because raised beds are an investment.

At the lower end, you can assemble a very basic bed for around fifty dollars if you are using simple materials or entry-level kits. These beds work. They are not decorative, but they do the job.

In the middle range, around one hundred fifty dollars, you start to see better materials, thicker boards, and designs that are more comfortable to work with. This is often the sweet spot for beginners who want something sturdy without overpaying.

Above that, you are mostly paying for aesthetics. High-end beds can look beautiful and last a long time, but they do not grow better vegetables. If you enjoy the look and it fits your budget, that is fine. Just do not assume it is necessary for success.

A common regret is spending heavily on the bed and then cutting corners on soil or plants. If you are choosing where to allocate funds, prioritize what goes inside the bed.

Height, Comfort, and Accessibility

Raised beds come in many heights, and taller is not always better.

Beds that are very tall can be helpful for people with mobility concerns or back pain, but they require significantly more soil. That increases cost and effort quickly.

Standard-height beds work well for most people and most vegetables. They allow you to sit or kneel nearby and reach in comfortably. They also make it easier to manage moisture and temperature in the soil.

If bending is a concern for you, consider a slightly taller bed rather than jumping straight to the tallest option available. Often, a modest increase in height provides enough comfort without the downsides of very deep beds.

Again, the goal is not to optimize. It is to choose something you will enjoy using.

What Not to Overthink

At this stage, it is tempting to worry about features that do not matter much yet.

You do not need built-in irrigation. You do not need elaborate liners. You do not need multiple tiers or compartments. These features can be useful later, but they are not required to grow vegetables.

Simple beds are easier to understand, easier to maintain, and easier to adjust as you learn. Complexity can always be added. Simplicity is harder to reclaim once you have committed to something complicated.

If a feature makes the bed feel intimidating or precious, it is probably not helping you.

Matching the Bed to Your Space

Before committing to a bed, it helps to visualize it in the space where it will live.

Measure the area. Mark the footprint on the ground if you can. Stand where you would water or harvest and see how it feels. This simple step prevents many regrets.

Make sure there is room to walk around the bed comfortably. Make sure it does not block access to other parts of your yard or patio. Make sure it does not feel like an obstacle.

A raised bed should feel like an invitation, not a barrier.

The Hidden Value of Modesty

There is a quiet advantage to starting with a modest setup.

Smaller beds are easier to fill, easier to plant, and easier to observe. They let you learn without pressure. They also make it easier to stop when you have done enough for the day.

Large, ambitious setups can be rewarding, but they demand more attention and more confidence. That can come later.

Starting small is not playing it safe. It is playing it smart.

What to Do Next

Once you have chosen the type and size of your raised bed, stop shopping and stop comparing. Decision fatigue is real, and it does not improve results.

If you are buying a bed, place the order and move on. If you are building one, gather the materials and prepare for the next step.

In the next chapter, we will walk through how to build a raised bed yourself, step by step, using simple designs and clear logic. Even if you plan to buy a bed, that chapter will help you understand what makes a sturdy, functional design.

Clarity comes from action, not from keeping options open.

Chapter 5:

Build Your Own Raised Bed

(Simple, Sturdy, No Carpentry Skills Required)

Two DIY options that are simple, sturdy, and realistic, even if your "toolbox" is currently a junk drawer

Before we get into wood types, measurements, or tools, it's important to say this out loud.

You do not have to build your own raised bed to grow vegetables successfully.

And if buying a kit keeps you moving instead of stalling, that is a smart choice, not a lazy one, but if you want the most bed for the least money, or you want a sturdier bed than most lightweight kits, or you simply like the idea of building something once and using it for years, DIY can be surprisingly doable when the design is simple and the instructions are written for a real human who does not build furniture for fun.

This chapter gives you two DIY options, both designed to be beginner-proof, both designed to avoid perfectionism traps, both designed so that if you can measure, screw things together, and tolerate the fact that wood is rarely perfectly straight, you can build a bed that grows food reliably.

Option 1 is the "I want this done today with minimal thinking" build, Option 2 is the "I want it sturdier and nicer without turning into carpentry class" build, and you can choose based on your energy, budget, and how much you want to rely on tools.

Beginner Shortcut

If you feel overwhelmed reading any DIY instructions, skip to **Option 1**, build it exactly as written, and move on. You can always upgrade later. Gardening rewards momentum far more than it rewards the most elegant construction.

Before You Build Anything: The Only Decisions That Matter

There are a hundred small decisions you could make, but most of them do not change the outcome in a way you will notice, so here are the only ones worth deciding upfront.

1) Bed size

For beginners, the most practical size is **4 feet wide**, because you can reach the middle from either side without stepping into the bed, and stepping into the bed is how soil becomes compacted and annoying.

Length is flexible, but **6 feet or 8 feet** is the sweet spot, long enough to be productive, short enough to build without wrestling boards like you are carrying a canoe.

Height depends on comfort and cost. A **10–12 inch** bed (one board high) is the simplest and cheapest, while a **20–24 inch** bed is more comfortable and holds moisture better, but uses more soil and more lumber.

If you are unsure, choose **4 ft x 8 ft x 12 in**. It works in most yards, it is productive, and it keeps costs reasonable.

2) Location

Build it as close as you reasonably can to water and to where you will actually walk, because the easiest garden to maintain is the one you bump into naturally.

3) How permanent you want it

Both builds here are "semi-permanent," meaning they can last years and still be moved if needed, but they are sturdy enough to feel stable and not like a flimsy frame.

Tools and Materials: What You Actually Need

You do not need a workshop. You need a short list and a willingness to keep things basic.

Basic tools for both options

- Tape measure
- Pencil or marker
- Drill or driver (corded is fine, cheap is fine)
- Bit for driving screws (usually Phillips or Torx)
- Work gloves
- Level (helpful but optional, a straight board and eyeballing works if needed)
- Safety glasses if you are cutting or drilling a lot

If you are cutting boards yourself

- A handsaw, circular saw, or miter saw. If you do not own a saw, do not panic. Many hardware stores will cut lumber for you if you give them your cut list, and if you build Option 1, you can often avoid cuts entirely.

Materials, simplified

You need boards for the walls, fasteners to hold corners together, and optionally something to reinforce long sides.

The big choice is wood type. Keep it simple:

- **Cedar**: lasts well, costs more, beginner-friendly choice if budget allows
- **Redwood**: great, often expensive, not always available
- **Untreated pine**: cheaper, does not last as long, still usable if you accept that you may replace boards sooner
- **Pressure-treated wood**: durable and commonly used; modern pressure-treated lumber sold for residential use is generally considered safe for garden structures when used correctly, but if the idea makes you uneasy, choose cedar and move on, because peace of mind matters too.

Reality Check

Wood moves. It swells and shrinks. It warps slightly. A bed can be a little imperfect and still be completely functional, so do not waste energy trying to make it look like a cabinet.

No Overthinking Rule #3

If the bed holds soil and does not wobble, it is good enough to grow food.

Option 1: The Simple Box Build

Fast, minimal cuts, minimal tools, strong enough for years

This build is the one you choose when you want a bed built quickly, cheaply, and without overengineering, and it is also the build that tends to work best for beginners because there are fewer opportunities to second-guess.

Recommended finished size

4 ft x 8 ft x 12 in (one board tall)

You can also do **4 ft x 6 ft x 12 in** if you want a smaller footprint.

What it looks like

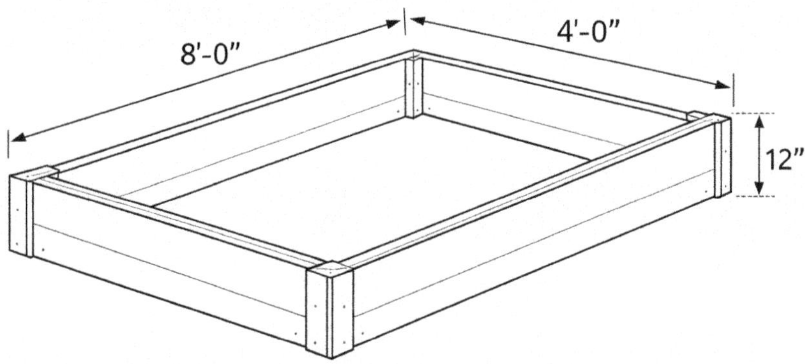

Side board meets end board at a right angle, screwed into a corner brace or corner post.

There are two beginner-friendly ways to do corners for Option 1. Pick the one that feels easiest.

Corner Method A: Metal corner brackets

This is the simplest if you want the least thinking.

You buy four heavy-duty corner brackets or raised-bed corner kits, attach boards to them, and you are done. This method is especially good if you cannot cut perfectly or you just want the corners to be strong without fiddling.

Heavy-duty corner bracket

Raised-bed corner kit

Corner Method B: 4x4 corner posts

This method is also easy, and it is very sturdy.

You cut four short 4x4 posts and screw your side boards into them. It is slightly more lumber, but it is forgiving and strong.

Cut list for Option 1 (4 ft x 8 ft x 12 in)

You are building a one-board-high rectangle.

Using 2x12 boards (recommended for 12-inch height)

- Two boards cut to **8 ft** (the long sides)
- Two boards cut to **4 ft** (the short ends)

If you buy 8-foot boards, you can keep the long sides full length and only cut the end boards.

If you choose 4 ft x 6 ft:

- Two boards at **6 ft**
- Two boards at **4 ft**

Corner posts method (optional)

- Four 4x4 posts cut to **12 in** tall. If you are stacking boards for a taller bed, you would cut taller posts, but Option 1 is meant to stay simple.

Hardware for Option 1

- Exterior-grade screws, **2.5 to 3 inches** long.

- If using metal brackets, follow bracket screw recommendations
- If using 4x4 posts, use **3-inch** screws for a strong bite
- Optional: two flat metal mending plates for each long side if your boards are slightly bowed and you want extra stability, not required but helpful

Beginner Shortcut

Buy a box labeled "deck screws" or "exterior screws." Indoor drywall screws snap and rust outside, and you will hate them later.

Step-by-step assembly for Option 1

Step 1: Lay out the shape where it will live

Place the boards on the ground in a rectangle, long boards on the sides, short boards on the ends, and stand back for a second to confirm you like the location.

This is the easiest time to adjust position, because once the bed is full of soil, moving it becomes a very different kind of project.

Step 2: Square it up without perfection

If you want a simple check, measure diagonally from one corner to the opposite corner, then measure the other diagonal. If the diagonals are close to the same, your bed is square enough.

Close counts here. Soil does not care if your rectangle is a tiny bit trapezoidal, and neither will your vegetables.

Step 3: Attach the first corner

Pick one corner and build it first, because once one corner is solid, the rest becomes easier.

If using metal corner brackets: Hold the bracket inside the corner, align the boards, and screw through bracket holes into the boards. Do not overthink it. If the boards are flush on top, you are fine.

If using 4x4 posts: Place the post inside the corner so the boards will screw into it, pre-drill a small hole if your wood is prone to splitting, then drive screws from the outside of the board into the post, two screws per board end is usually enough.

Step 1: Lay out the shape where it will live

Place the boards on the ground in a rectangle, long boards on the sides, short boards on the ends, and stand back for a second to confirm you like the location.

Step 2: Square it up roughly

Measure diagonally from one corner to the other. If the diagonals are close to the same, your bed is square enough.

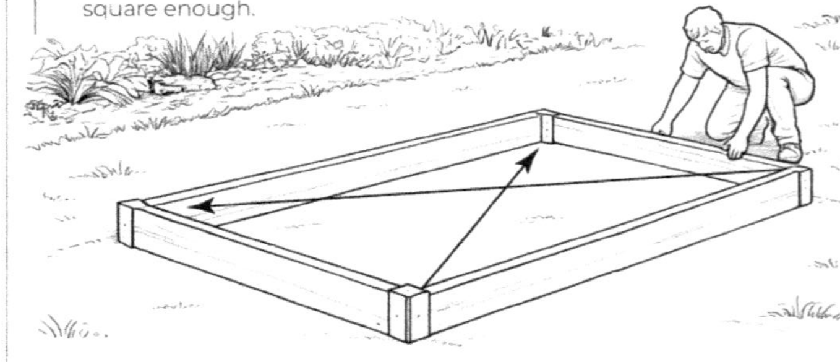

Step 3: Attach the first corner

Pick one corner and build it first, because once one corner is solid, the rest becomes easier.

If using metal corner brackets:

Hold the bracket inside the corner, align the boards, and screw through bracket holes into the boards.

Step 4: Attach the opposite corner

Keeping one long board attached, move to the other end of that same board and attach the second corner.

If using 4x4 posts:

Place the post inside the corner so the boards will screw into it, pre-drill a small hole if your wood is prone to splitting, then drive screws from the outside of the board into the post, two screws per board end is usually enough.

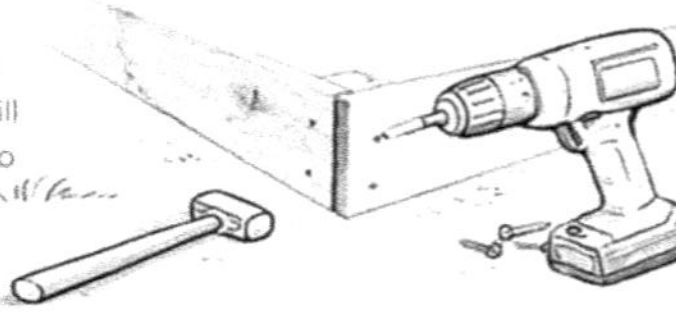

Step 4: Attach the opposite corner on the same side

Keeping one long board attached, move to the other end of that board and attach the second corner. This keeps the long side straight and reduces wobble.

Step 5: Attach the remaining corners

Once two corners are done, the frame holds its shape better, and the remaining corners are straightforward.

Step 6: Reinforce long sides if needed

If your long boards bow outward slightly, add a simple reinforcement.

The easiest reinforcement is a center stake or brace, such as a short 2x2 or 2x4 placed vertically inside the bed at the midpoint of each long side, screwed into the board to keep it straight.

For a one-board-high bed, this is usually optional. For taller beds, reinforcement becomes more important, and Option 2 handles that better.

Step 7: Place the bed and level it enough

You do not need laser-level perfection.

If the ground slopes slightly, soil will settle, and plants will still grow. What you want to avoid is a dramatic tilt that causes water to run to one corner.

If needed, scrape away high spots, add a little soil under low spots, and aim for "pretty level," not "perfect."

Ground contact, liners, and weed barriers for Option 1

This is where beginners often overcomplicate things, so here is the simplest truth.

Do you need a weed barrier under the bed

If the bed sits on grass or weeds, lay down **plain cardboard** in overlapping sheets and wet it thoroughly before adding soil. It suppresses weeds and breaks down over time, and it does not create drainage issues the way some fabrics can.

Do you need to line the inside with plastic

No, and most of the time you should not. Plastic traps moisture against the wood and can shorten the life of the boards, and it is also unnecessary for plant health.

What about hardware cloth for burrowing pests

If you have a known problem with gophers, moles, or other burrowing pests, attaching hardware cloth to the bottom before placing the bed can help. If you don't know you have that problem, you do not need to solve it preemptively.

Reality Check

The most profitable garden is not the one protected from every hypothetical issue, it's the one you actually build and plant.

Option 2: The Sturdier Stacked Build

More comfortable height, stronger walls, still beginner-friendly

Option 2 is for you if you want a taller bed for comfort, deeper soil volume for steadier moisture, or simply a structure that feels solid and "done," the kind of bed you can keep for years without it bowing or wobbling, and the good news is that sturdier does not have to mean complicated.

This build uses two layers of boards stacked, corner posts for strength, and simple mid-span bracing so the long sides do not bulge under soil pressure.

Recommended finished size

4 ft x 8 ft x 22–24 in tall

You can reach the middle easily, you get more soil depth, and the bed is comfortable to work around without kneeling constantly.

What it looks like

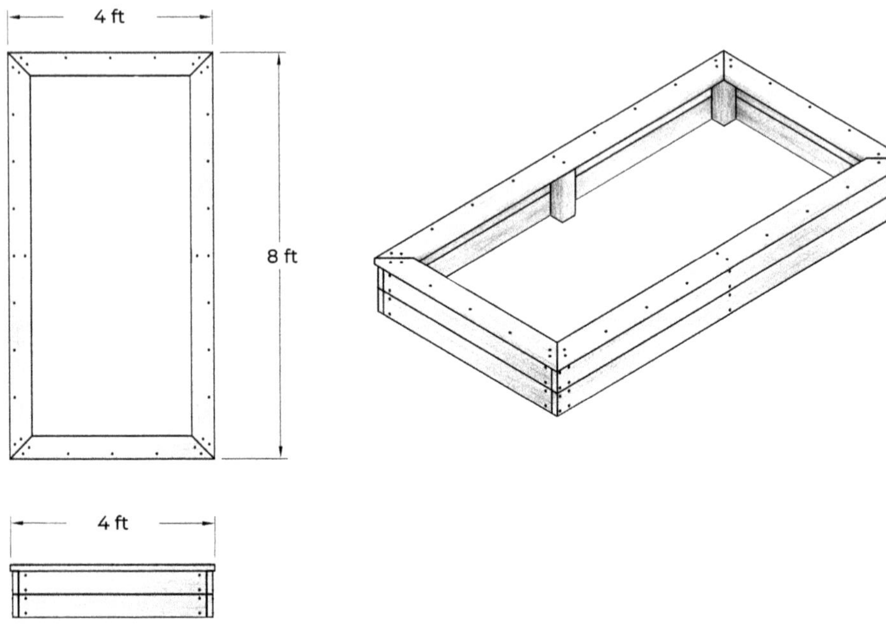

Corner posts inside, braces at midpoint

Cut list for Option 2 (4 ft x 8 ft x 24 in)

Boards

Using 2x12 boards again:

- Four boards at **8 ft** (two per long side, stacked)
- Four boards at **4 ft** (two per short end, stacked)

That is eight boards total for the walls.

Corner posts

- Four 4x4 posts cut to **24 in** (or 22 in if you want slightly shorter)

Mid-span bracing

Choose one of these approaches.

Brace Method A: Vertical inside stakes (simplest)

- Two 2x4 stakes cut to **24 in** for the midpoint of each long side, placed inside the bed

Brace Method B: Through-bolted tie (more advanced but very sturdy)

- One or two cross ties (a board or threaded rod) spanning across the width, preventing the long sides from pushing outward
  This is optional, and most beginners do not need it for a 4 ft wide bed if vertical bracing is used.

Hardware for Option 2

- Exterior screws, **3 inches** for attaching boards to 4x4 posts
- Exterior screws, **2.5 inches** for attaching stacked boards to each other where needed
- Optional: washers and bolts if you want to bolt instead of screw, not required
- Optional: corner brackets for extra rigidity, but corner posts alone are usually enough

Beginner Shortcut

Pre-drill holes near board ends, especially if your wood splits easily, because pre-drilling turns "I can't do this" into "Oh, that was fine."

Step-by-step assembly for Option 2

Step 1: Build one long wall first

Lay one 8-foot board flat and place a 4x4 post at each end, aligned flush with the board's top edge, then screw the board into the post with two to three screws spaced vertically.

Now place the second 8-foot board directly above or below it to create a stacked wall, align the edges as best you can, and screw it into the same posts.

You now have one complete long wall, two boards high, anchored by posts.

Step 2: Build the second long wall

Repeat the same process. Keep the posts oriented the same way so assembly is straightforward when you bring everything together.

Step 3: Connect the short ends

Stand the two long walls upright, parallel to each other, spaced about 4 feet apart, then attach the 4-foot boards to connect them, one bottom layer board first, then the top layer board.

If you are working alone, this is the moment where it feels like you need a second person, so here is the trick: use something heavy, like a brick, a bucket of soil, or even a chair, to hold one wall steady while you attach the end boards, and do one corner at a time rather than trying to align everything at once.

Step 1 & 2: Build long walls

Lay one 8-foot board flat and place a 4x4 post at each end, aligned flush with the board's top edge, then screw the board into the post with two to three screws spaced vertically. Repeat.

Step 3: Connect the short ends

Stand the two long walls upright, parallel to each other, then attach the 4-foot boards to connect them, one bottom layer board first, then the top layer board.

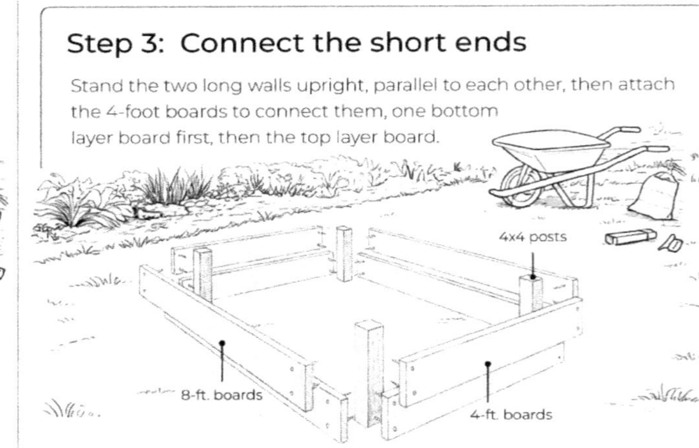

Step 4 & 5: Check and add mid-span bracing

Check the diagonals (see Step 2 for Build Option 1). Place a vertical 2x4 stake inside the bed at the midpoint of each long side, press it flush against the boards, and screw through the board into the stake.

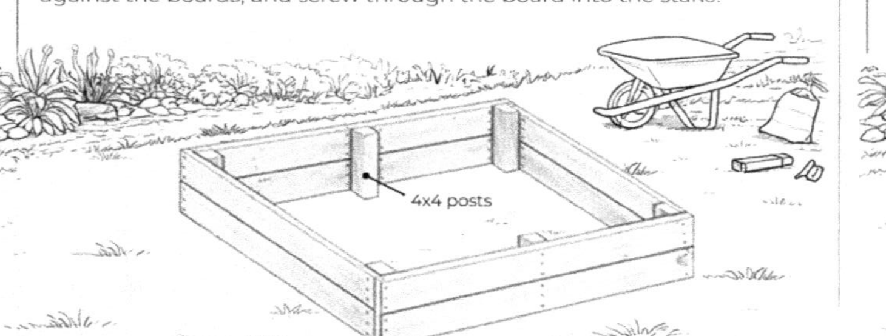

Step 6: Decide whether you want a top cap

If you do, use 2x4 boards laid flat along the top edges and screw them into the walls.

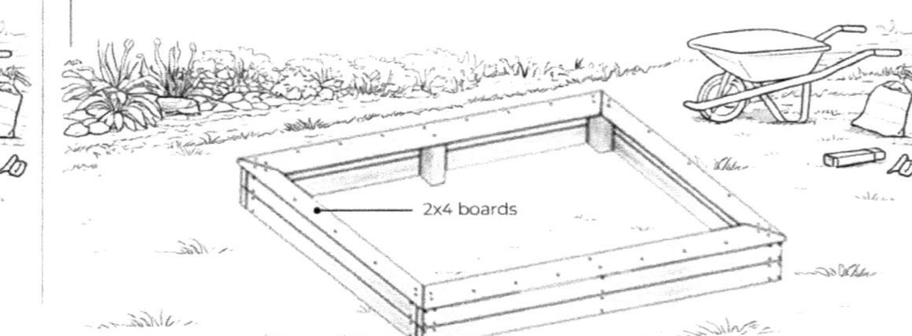

Measure diagonals if you want, but remember, you are building a garden bed, not a piano. If it looks square and the corners meet cleanly enough, you are good.

Step 5: Add mid-span bracing

For a bed this tall, soil pressure can push long boards outward over time, so bracing is worth doing, and it is not hard.

Place a vertical 2x4 stake inside the bed at the midpoint of each long side, press it flush against the boards, and screw through the board into the stake. This creates a strong "rib" that keeps the wall straight.

If you want extra rigidity, you can also add a second stake on each long side, roughly at one-third and two-thirds of the length, but for most 4x8 beds, one brace per long side is a solid start.

Step 6: Decide whether you want a top cap

A top cap is a flat board placed along the top edge like a rim. It is optional, but it can make the bed more comfortable to lean on, it can stiffen the structure, and it can make the bed look more finished.

If you do not care, skip it.

Soil containment and longevity tips for Option 2

Taller beds last longer and behave better when you prevent water from sitting against the wood unnecessarily.

- Make sure the bed drains well and is not sitting in a low, swampy spot

- Avoid lining with plastic
- Use mulch inside the bed to reduce soil splash and moisture extremes
- Accept that wood is a consumable material, even cedar eventually ages, and that is normal

---

"Fluffy Slippers" Safety and Sanity Notes

This section is not dramatic, it's practical.

- If you are cutting wood, wear eye protection, because sawdust in the eye will ruin your entire mood and probably your day
- If a board is heavy or awkward, do not muscle it, rest one end on something and move it in stages
- If you miss a screw hole and the screw goes in crooked, back it out and try again, you did not "ruin it"
- If your boards are slightly bowed, install them with the bow facing inward, soil pressure often pushes them out and makes them straighter over time
- If you have never used a drill, practice on scrap wood first, you will be surprised how quickly it feels normal

---

Troubleshooting Common DIY Issues

"My boards don't line up perfectly"

That is normal. Focus on corners being secure and top edges being reasonably level. Small gaps do not matter.

"The bed rocks a little"

Ground is uneven. Scrape a high spot, add soil under a low spot, and aim for stable enough that it doesn't wobble when touched.

"The long sides are bowing outward"

Add a mid-span brace inside. This is the correct fix and it's easy.

"I built it, and now I hate the location"

Move it now, empty, before soil goes in. If soil is already in, remove soil into containers, move the bed, refill, and treat it as a one-time annoyance, not a life sentence.

Quick Comparison: Option 1 vs Option 2

Option 1 is best when you want speed, simplicity, and a low-cost starting point, and you are fine with a lower bed height.

Option 2 is best when you want comfort, longer-term sturdiness, and less bending, and you are willing to spend more on lumber and soil.

Both grow vegetables. Neither requires perfection.

What to Do Next

Once your bed is built and placed, do not get stuck in the finishing-details loop, where you start thinking you need to line it, seal it, paint it, decorate it, or "complete" it before it deserves soil, because the only finishing step that matters is filling it and planting it.

In the next chapter, we'll make soil simple, including exactly what to buy, how much you need, and what to ignore, so you can get from empty bed to planted bed without turning it into a research project.

For now, the win is this: you built the container, you created the structure that reduces confusion and increases forgiveness, and once soil is in, you are no longer preparing to garden, you are gardening.

Chapter 6:

Soil Made Simple

(No Chemistry Degree Required)

How to fill your raised bed in a way that works, lasts, and does not turn into an obsession

Soil is where many people quietly lose confidence in gardening, not because it is particularly difficult to work with, but because it is talked about as though it were a fragile system that can be ruined by a single wrong choice, when in reality soil, especially in a raised bed, is far more tolerant and forgiving than most advice suggests.

If you have ever found yourself standing in a garden center, staring at stacks of soil bags with increasingly vague labels, unsure whether the difference between "premium," "organic," "vegetable," or "professional blend" mattered enough to justify the price difference, this chapter exists to remove that pressure and replace it with a small number of decisions you can make once and then stop revisiting.

The goal here is not to teach you everything there is to know about soil. It is to help you fill your raised bed with something that works well enough that plants grow, problems stay manageable, and your attention can move on to planting and harvesting, where gardening actually starts to feel rewarding.

No Overthinking Rule #4

Do not buy amendments before you have soil in the bed and plants in the ground.

Why Soil Advice Feels So Intimidating

Soil anxiety usually comes from the same two sources, even if they are not obvious at first.

The first is marketing, where every product is positioned as essential, optimized, or superior in some subtle way, creating the impression that success depends on choosing correctly among dozens of nearly identical options.

The second is overeducation, where beginners are introduced too early to concepts like nutrient ratios, microbial activity, and pH balance, without first being given permission to succeed with soil that is simply functional.

Raised beds change the starting conditions dramatically. You are not fighting compacted ground, unknown subsoil, or decades of neglect. You are creating a contained environment with good drainage and defined boundaries, which means your soil does not have to overcome nearly as much as it would in an in-ground garden.

Once you understand this, soil stops being the thing that determines whether you succeed or fail and becomes simply the medium plants grow in while you learn.

The Emotional Weight We Attach to Soil (And Why That's Unfair)

Over time, this constant emphasis on getting soil "right" does more than confuse people. It quietly shifts responsibility for every outcome onto the soil itself, which is where the emotional weight really starts to build.

Soil ends up carrying the emotional weight of the entire garden, because when plants struggle, growth slows, or enthusiasm fades, soil is often the first and only thing blamed.

What rarely gets acknowledged is that soil is not a static ingredient you choose once and then live with forever. It is a living, shifting medium that responds to water, roots, weather, and time, and in raised beds especially, it is designed to improve rather than degrade as you use it.

Once you let go of the idea that soil has to be "right" before you plant anything, you free yourself to see it for what it actually is, a starting point rather than a verdict.

What Soil Actually Needs to Do (And What It Doesn't)

Good raised bed soil has a very short job description.

It needs to hold moisture long enough for roots to access it, drain excess water so roots can breathe, provide physical support so plants stand upright, and supply nutrients at a pace plants can actually use.

That's it.

It does not need to be sterile, perfectly balanced, or chemically precise. It does not need to look rich and crumbly in every handful, and it certainly does not need to impress anyone.

Plants evolved to grow in far less controlled environments than your raised bed, and they are surprisingly adaptable when basic conditions are met.

What Plants Forgive Easily (And What They Don't)

Plants are remarkably forgiving of certain imperfections and far less forgiving of others, and knowing the difference helps you stop worrying about the wrong things.

They forgive uneven texture, minor nutrient imbalances, and soil that settles or looks slightly different from one corner to another.

What they struggle with is soil that stays waterlogged, soil that dries into a brick, or soil that never holds moisture long enough for roots to access it.

This distinction matters because many beginners spend energy correcting things that plants barely notice, while missing the few structural issues that actually affect growth.

Raised beds already eliminate most of the unforgivable problems, which is why soil choice in this context does not need to be perfect to be effective.

The Simplest Soil Decision That Works for Most People

If you want the least stressful path, start with a commercial raised bed or garden soil blend and stop there.

This is not because homemade mixes are wrong, but because they add decisions early on without delivering benefits most beginners will notice.

A bag labeled "raised bed soil," "garden soil," or "vegetable garden mix" is designed to do exactly what you need it to do, even if the brands differ slightly in texture or composition.

You do not need the most expensive option or the most amended option. You need soil that drains well and does not smell sour or rotten.

If you can water it and see moisture soak in without puddling or instantly disappearing, you are already in a good place.

When DIY Soil Mixes Make Sense (And When They Don't)

At some point, you will encounter recipes for DIY soil mixes, often presented as universal formulas that promise better drainage, better nutrients, or better biology.

While these mixes can work well, they are rarely necessary at the beginning.

DIY mixes make the most sense when you already understand how your bed behaves, how quickly it dries, how compact it becomes, and how your plants respond over time.

For a first season, DIY mixes often add complexity without delivering clarity, because when something goes wrong, you have more variables to question and fewer reference points to work from.

Starting with a commercial mix gives you a known baseline, which makes observation and adjustment easier later, when you have context for what your garden actually needs.

How Much Soil You Really Need (And Why People Underestimate It)

One of the most common practical frustrations with raised beds is underestimating soil volume, which leads to half-filled beds, uneven settling, and unnecessary second trips to the store.

A standard four foot by eight foot raised bed that is twelve inches deep requires roughly thirty-two cubic feet of soil, which typically translates to around eighteen to twenty-two large bags, depending on the brand and bag size.

A taller bed, around twenty-four inches deep, needs roughly double that volume, but that does not mean you need to fill the entire bed with premium soil.

If the numbers feel abstract, the simplest approach is to ask the garden center how many bags they recommend for your bed size, then buy a couple extra. Extra soil is easier to store or use later than trying to stretch what you have to fill a bed properly.

Filling a Tall Bed Without Overspending

For deeper beds, it makes sense to think in layers, not because plants require it, but because your budget will appreciate it.

The bottom third to half of a tall bed can be filled with lower-cost material that still drains well, such as basic garden soil, soil mixed with compost, or partially decomposed organic matter, as long as it is not compacted or waterlogged.

The top ten to twelve inches is where most plant roots spend their active time, and that is where higher-quality soil matters most.

This layered approach mirrors how soil naturally builds over time and does not compromise plant health when done sensibly.

A Slow, Literal Walkthrough of Filling the Bed

Once the bed is built and placed, filling it can feel oddly anticlimactic and strangely stressful at the same time, especially if you are worried about doing it "wrong," so it helps to slow this process down and treat it as a sequence rather than a single task.

Start by placing any cardboard or weed suppression layer on the ground beneath the bed, overlapping pieces slightly and wetting them so they stay in place.

If you are using lower-cost fill for the bottom portion, add it first and spread it evenly without packing it down aggressively. Light settling is fine. Compression is not necessary.

Next, add your primary planting soil to the top portion of the bed, spreading it loosely and watering as you go, which helps eliminate large air pockets without turning the soil into mud.

Expect the soil level to drop slightly after watering. This is normal and does not mean you miscalculated.

Resist the urge to top it off repeatedly in one session. Let it settle overnight if possible, then assess whether you truly need more.

This slower approach reduces anxiety and prevents overfilling, which can cause soil to spill over the sides or compact unnecessarily.

Compost, Explained Without the Mythology

At some point, compost enters the conversation, usually framed as the thing that will finally make everything work.

Compost is simply decomposed organic matter. It improves soil structure and adds nutrients slowly.

A reasonable amount of compost in raised bed soil is helpful. Too much can cause problems.

As a general guideline, compost should make up about one quarter of the soil volume in the planting zone. More than that can lead to excessive nutrients, poor drainage, or soil that collapses as organic matter continues to break down.

If you are unsure, use less. You can always add compost later as a top dressing.

A Walk Through the Garden Store, Mentally

When you walk into a garden center, the goal is not to decode every label, but to avoid obvious red flags.

Pick up a bag and feel it. It should feel light, not dense. If possible, smell it. It should smell earthy, not sour or swampy.

Ignore marketing words and look for function. Raised bed soil should drain when watered and hold together lightly when squeezed, then break apart easily.

If you find one or two options that meet those criteria, choose the one that fits your budget and move on.

The longer you stand there comparing labels, the more likely you are to second-guess a decision that would have worked perfectly well.

What Bad Soil Looks Like in Real Life

Bad soil is not subtle, and recognizing it is easier than diagnosing nutrient charts.

It often feels heavy and sticky when wet, or dusty and water-repellent when dry. Water may pool on the surface or run straight through without being absorbed. Plants may wilt even when the soil is wet or grow weak and pale despite regular care.

If you encounter these signs, it is not a personal failure. It simply means the soil structure needs improvement, which is fixable over time.

What "Okay" Soil Looks Like (Which Is What You're Aiming For)

Okay soil does not look impressive in photos.

It may look uneven, contain small pieces of organic matter, or feel slightly different from one handful to the next.

Okay soil drains when watered, holds moisture for a reasonable amount of time, and supports plants upright without collapsing around the stems.

If your soil meets those criteria, it is doing its job, even if it does not match the idealized images you have seen online.

This matters because chasing visual perfection in soil often leads to unnecessary amendments that disrupt what was already working well enough.

Why You Do Not Need to Test Soil in Your First Season

Soil testing can be useful, but it is rarely necessary in the first season of a raised bed garden.

Most beginner issues are related to watering patterns, plant choice, or impatience, not nutrient deficiencies that require laboratory confirmation.

Testing too early often leads to overcorrection, where amendments are added that create new problems rather than solving old ones.

If plants are growing reasonably well, testing can wait.

How Early Testing Can Actually Slow Learning

Testing soil too early often shifts attention away from observation and toward correction, which can be counterproductive when you are still learning how plants respond to basic care.

Numbers without context do not tell you what to do next. They invite action without understanding, which can lead to adding nutrients, adjusting pH, or changing watering patterns unnecessarily, creating new variables before old ones have been understood.

Learning how soil behaves through direct observation builds intuition faster than interpreting test results without experience to ground them.

How Soil Improves Without You Doing Much

One of the most reassuring truths about soil is that it improves with use.

As roots grow and die back, as organic matter breaks down, and as microbes establish themselves, soil structure becomes more stable and forgiving, especially when you are not constantly disturbing it.

Raised beds benefit from this process more than in-ground gardens because conditions are controlled and compaction is minimal.

Each season you garden, the soil becomes easier to work with, not harder.

Why Leaving Soil Alone Is Often the Best Choice

One of the hardest habits for beginners to break is the urge to constantly improve soil, tweak it, adjust it, or "help" it along, when in reality stability is what allows soil systems to develop.

Every time soil is heavily disturbed, it resets moisture patterns and microbial activity, which can slow the very improvements you are trying to encourage.

Raised beds thrive when soil is allowed to settle into a rhythm, supported by mulch and consistent watering, rather than frequent intervention.

Doing less, in this case, produces more reliable results.

Mulch: The Quiet Helper Most People Skip

Mulch deserves more credit than it gets.

A layer of mulch protects soil from temperature extremes, reduces moisture loss, and prevents surface compaction. It also reduces how often you need to water and makes soil behavior more predictable.

Mulch does not need to be decorative. Straw, shredded leaves, or bark fines all work.

A few inches is enough to make a noticeable difference.

Common Soil Mistakes That Look Like Good Ideas

Adding fertilizer immediately, before plants show any need for it, often causes more harm than good.

Replacing soil every season is unnecessary and wasteful.

Obsessing over uniform texture is counterproductive.

Trying to “fix” soil daily prevents it from stabilizing.

Soil rewards patience far more than intervention.

First Season Soil Expectations vs Reality

In the first season, soil will settle. It will look lower after watering. It may feel uneven. This is normal.

Do not top it up repeatedly out of anxiety. Let it settle, mulch it, and plant.

By the second season, soil will feel more cohesive and easier to work with, and you will notice fewer extremes in moisture and temperature.

This progression is a sign the system is working.

The Long View: Soil Is a Relationship, Not a Setup Step

It helps to think of soil not as a setup task you complete and move on from, but as a relationship that develops gradually, season by season, shaped by what you grow, how you water, and how often you disturb it.

The first season is about establishing structure and learning basic behavior.

The second season is about refinement.

Later seasons are where soil starts to feel predictably supportive rather than mysterious.

Seeing soil through this lens removes pressure from the first fill and replaces it with patience, which is far more useful.

What to Do Next

Once your bed is filled, water the soil thoroughly and let it settle for a day or two before planting.

Resist the urge to keep adjusting or improving it.

In the next chapter, we will focus on planting itself, including spacing and depth, so you can move from preparation to action without carrying unnecessary doubt forward.

For now, trust this: if your soil drains, holds moisture, and supports plants upright, it is doing its job well enough, and that is all it needs to do.

PART III:

Plant Without Guesswork

Chapter 7:

Planting That Actually Works

This is the part many people think they should feel confident about by now, and yet somehow don't.

They have a raised bed. It's in a decent spot. It's filled with good soil.

And still, when it comes time to plant, hesitation creeps in. How deep is too deep. How close is too close. Whether seeds are better than seedlings. Whether now is the right time or whether waiting another week would be smarter.

If that sounds familiar, you're not behind. This is exactly where uncertainty tends to show up.

Planting feels permanent in a way the earlier steps did not. Once something is in the ground, it feels like you've committed. That's why this chapter focuses less on rules and more on removing pressure.

You don't need to plant perfectly. You need to plant in a way that gives your vegetables a fair start.

That is much easier than it sounds.

Seeds or Seedlings: Which Is Easier, Really

This question causes more stress than it deserves.

Seeds and seedlings can both work well. The difference is not about skill. It's about patience, timing, and how much uncertainty you're comfortable with at the beginning.

Seedlings are often easier for beginners. They give you an immediate sense that something is happening. You can see leaves. You can tell if a plant looks healthy. You skip the delicate germination stage, which is where many first-time gardeners worry they did something wrong.

Seeds are less expensive and give you more variety, but they require a little faith. There is a quiet stretch where nothing is visible, and that can be uncomfortable if you're already unsure.

If you want the calmest start possible, use seedlings for your first bed. You can always experiment with seeds later, once confidence has settled in.

If you enjoy the idea of starting from the very beginning and you're okay with waiting a bit longer for visible progress, seeds are fine too.

There is no wrong choice here.

A Simple Way to Decide Without Overthinking

If you're unsure which route to take, use this rule.

Choose seedlings for anything you want results from quickly or that feels important to you. Choose seeds for fast-growing greens where a delay of a week or two doesn't matter much.

That way, you hedge your bets. You get early wins and still learn how seeds behave in your space.

When to Plant, Without a Calendar Obsession

Gardening advice often comes with planting calendars, frost dates, and zone charts. Those tools can be useful, but they also make beginners feel like they're constantly late or early.

Here's a calmer way to think about timing.

Most vegetables are looking for stable conditions, not exact dates. They want temperatures that are not extreme and soil that is workable. In many parts of the US, that window is wider than you think.

If local nurseries are selling a plant, that is usually a sign it's an acceptable time to plant it. Nurseries move inventory based on local conditions, not theory.

For seeds, the package will usually give a general temperature range. You don't need to hit it perfectly. Close is good enough.

If you plant a little early and growth is slow, plants often catch up once conditions improve. If you plant a little late, many beginner-friendly vegetables still perform well.

The bigger mistake is waiting so long that nothing gets planted at all.

How Deep to Plant Without Memorizing Charts

Planting depth sounds technical, but it's surprisingly intuitive once you understand the principle.

Seeds are planted shallowly. Seedlings are planted at the depth they were already growing.

For seeds, a common guideline is to plant them about **twice as deep as their size**. Tiny seeds barely get covered. Larger seeds go a bit deeper. You do not need a ruler. You need reasonable judgment.

For seedlings, place them so the top of the root ball sits level with the surrounding soil. You're not burying the stem unless the plant specifically benefits from it, which most beginner vegetables do not.

If this feels vague, that's because it is meant to be. Plants evolved without measuring tools. They are adaptable.

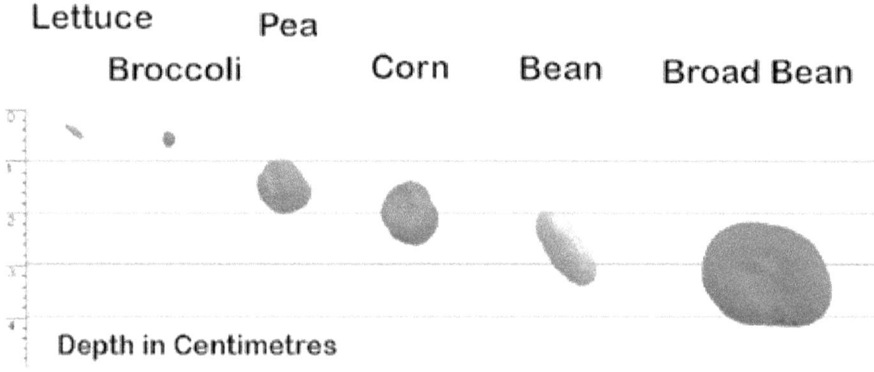

These visuals show the general idea. You are aiming for supportive contact with the soil, not precision.

Spacing: Where People Try Too Hard

Spacing is another area where beginners often make things harder than they need to be.

Seed packets and plant labels usually list spacing recommendations. Those numbers are useful, but they are not laws. They assume ideal conditions and full-sized plants.

In a raised bed, slightly closer spacing often works well, especially for leafy greens. Plants support each other, shade the soil, and reduce moisture loss.

The real problem comes from crowding fruiting plants or packing everything tightly because empty space feels wasteful.

Empty space is not waste. It's airflow, flexibility, and room to learn.

A simple approach is to space plants so their mature leaves can touch without overlapping heavily. If you imagine the plant at full size and give it that footprint, you're close enough.

If you're unsure, err on the side of a little more space. You can always add more later. You can't easily remove plants without disturbing roots.

Planting Layouts That Reduce Stress

For your first raised bed, avoid complicated layouts.

Straight rows are fine. Loose groupings are fine. What matters is that you can tell which plant is which and that each one has room to grow.

Many beginners enjoy planting in simple blocks, grouping the same plant together. This makes watering, harvesting, and observation easier.

Mixed planting also works, especially with herbs and greens, but it's not required to get started.

The more clearly you can see what's happening in your bed, the easier it is to learn.

How to Handle the "Did I Do This Right" Moment

Almost everyone has this moment right after planting.

You step back. You look at the bed. It feels underwhelming. The soil looks bare. The plants look small. You wonder whether you should add more or change something.

This is normal.

Plants do not look impressive on planting day. Growth happens quietly at first. Resist the urge to fix what is not broken.

Water gently. Walk away. Check back in a few days.

Gardening rewards restraint more often than intervention.

Watering Right After Planting

Newly planted seeds and seedlings need moisture, but not saturation.

The goal is to settle the soil around roots or seeds so there are no air gaps. A gentle, thorough watering accomplishes this.

Avoid blasting the soil or washing seeds away. If you're worried about disturbance, use a watering can or a hose with a gentle setting.

After that first watering, let the soil surface dry slightly before watering again. Constant wetness is more harmful than brief dryness.

What to Expect in the First Two Weeks

This is a helpful window to understand.

In the first week, seeds may show no visible change. Seedlings may look unchanged. This does not mean nothing is happening. Roots are adjusting before top growth accelerates.

In the second week, you often see new leaves or clearer growth. This is when confidence usually starts to build.

If nothing has emerged from seeds after a reasonable time, that's information, not failure. You can replant. Seeds are inexpensive. The bed is still usable.

This is one of the advantages of starting with forgiving plants.

Common Early Mistakes That Are Easy to Fix

Planting too deep, watering too often, and crowding plants are very common early mistakes. The good news is that they are usually fixable.

Plants tell you when they're unhappy. Leaves change color. Growth slows. These signals are not judgments. They are feedback.

If something looks off, pause before reacting. Often the best adjustment is a small one.

Letting the Bed Teach You

Every raised bed has its own rhythm. Light, heat, wind, and moisture behave slightly differently in each space.

The first planting is as much about observation as it is about harvest. Notice which areas dry faster. Notice which plants respond quickly. Notice what feels easy.

You don't need to document everything. Just pay attention.

This quiet learning phase is where gardening starts to feel less intimidating and more familiar.

What to Do Next

You've planted everything. The soil is watered. The bed looks… quiet.

This is usually the moment people start wondering if they did something wrong.

Planting day often feels underwhelming. There is no big visual payoff. The bed doesn't look "full" yet. Seeds are invisible. Seedlings are small. It can feel like all that effort disappeared into the soil.

That feeling is normal.

What happens next is not dramatic or obvious. For a while, it's mostly invisible. Roots adjust. Seeds settle. Plants pause before they move again. This is not a problem. It's the process.

The next phase is where most beginners lose confidence, not because something is actually going wrong, but because nothing seems to be happening yet.

That's why the first two weeks deserve their own attention. They come with a specific kind of uncertainty, and knowing what's normal during this window makes everything that follows much easier.

In the next chapter, we'll slow things down and walk through exactly what to expect in those first fourteen days. What's normal. What isn't. When to wait. When to intervene. And when doing nothing at all is the right move.

For now, you haven't missed anything. You've done enough.

Chapter 8:

The First 14 Days:

What's Normal, What's Not

Why This Quiet Stretch Matters More Than You Think

The first two weeks after planting are rarely dramatic, and that is precisely why they feel so uncomfortable for people who are new to raised bed gardening, because by this point you have already done the visible work. You built the bed, filled it with soil, planted seeds or seedlings, watered carefully, and stepped back with the quiet hope that something would happen quickly enough to confirm that you did it right.

Instead, what usually happens is very little, at least on the surface, and the soil looks calm, the bed looks unfinished, and the plants, if they are visible at all, look smaller and less confident than the versions you imagined while planting. It is in this gap between effort and reward that doubt tends to settle in, not loudly, but persistently.

This chapter exists to slow that moment down, not to rush you through it or distract you from it, but to explain what is happening, what is not happening, and what absolutely does not need fixing yet, so that you can move through this early phase without turning the garden into a source of stress.

The Morning After Planting

For many people, the first real wave of uncertainty arrives the morning after planting, once the initial momentum wears off and the bed has had time to look disappointingly unchanged.

You wake up, maybe earlier than usual, and the garden is one of the first things you think about. Before coffee, before checking your phone, you step outside or look out the window, half expecting the bed to look different somehow, as if the act of planting should have triggered visible change overnight.

Of course, it hasn't.

The soil looks the same, the bed looks quiet, and if you planted seeds, there is nothing to see at all. If you planted seedlings, they may look a little tired, slightly slumped, or unchanged from the day before, which is often when people feel a small drop in mood they don't quite name.

Yesterday felt productive and hopeful. Today feels flat. The contrast can be surprising.

It is also completely normal.

Plants do not respond to planting the way humans respond to completing a task. They do not rush to show appreciation. They pause, adjust, and redirect energy into places you cannot see yet, and that pause can feel like failure if you don't know to expect it.

This is also when the urge to intervene first appears, whether that means watering again even though you watered thoroughly the day before, touching the soil just to feel that something is happening, or wondering whether you should have planted deeper, spaced differently, or chosen something else entirely.

None of those urges mean you made a mistake. They mean you care, and caring without feedback is uncomfortable.

Why Waiting Feels So Hard at the Beginning

Once that first morning passes, the discomfort doesn't disappear. It usually changes shape.

Most modern tasks reward speed, clarity, and visible progress. You send an email and get a reply. You complete a checklist and see boxes ticked. You clean a space and immediately see the result of your effort.

Gardening does not work like that, especially in the early days.

In the first two weeks, almost everything that matters is happening below the surface, where seeds are absorbing moisture, roots are stretching into new soil, and plants are quietly deciding how to orient themselves for what comes next. Because you cannot see this work, your brain fills the silence with questions.

Did I water enough?

Did I water too much?

Did I plant too deep?

Should something be sprouting by now?

These questions are not signs that something is wrong. They are signs that your expectations are still calibrated to faster systems.

Learning to garden involves learning to tolerate this slower rhythm, and the first two weeks are where that learning begins.

Days One to Three: The Bed That Looks Empty

During the first few days after planting, raised beds often look disappointingly unchanged, which can be especially unsettling if you were expecting at least some sign that things are underway.

Seeds are underground, doing invisible work. Seedlings often pause after transplanting, redirecting energy into their roots before committing to new leaf growth, and during that pause they may look unimpressive or even slightly worse than when you planted them.

This is normal.

What makes it difficult is that your role has suddenly shifted. Up until now, you were actively building and doing. Now, your primary task is to wait, observe, and resist the urge to interfere, which is not a skill most people have practiced.

This is where many beginners start second-guessing themselves. They lean over the bed, gently scrape at the soil to check seed depth, or water again "just in case," even though the soil is already moist.

If there is one thing to practice during these first few days, it is restraint.

Water enough to keep the soil evenly moist, then step back. Do not dig for seeds. Do not reposition seedlings unless they are clearly collapsing. Trust that the work is happening, even if it is invisible, because nothing visible yet does not mean nothing is happening.

A small, very normal thing also happens in these first few days that almost nobody warns you about, which is that you start interpreting the bed the way you would interpret a work project. If you can't see progress, you assume the system is stalled, and then your brain starts searching for a reason, and because seeds are invisible, the only thing left to blame is you.

It might sound silly written down, but it's real in the moment. You find yourself thinking that maybe you should have bought seedlings instead, maybe your soil isn't right, maybe you planted too deep, maybe the sun is wrong, maybe you should have started earlier in the season, or maybe you're just not the kind of person who can keep plants alive. None of that is gardening reality. It's just your mind trying to close the loop between effort and reward.

This is also why the "just check once a day" advice matters, because it interrupts the mental spiral. When you check too often, you don't actually gain information, you just rehearse uncertainty repeatedly until it starts to feel like proof.

So if you want a simple rule for days one to three, use this: water only if the soil under the surface is drying out, and otherwise treat the bed like a slow cooker, not a microwave. Nothing looks different for a while, then suddenly it does, and you don't get better results by lifting the lid every ten minutes.

Days Four to Seven: The Week of Doubt

As the first week progresses, uncertainty often shifts into something more specific.

Some seeds may have emerged while others have not. One corner of the bed may look more active than another. Seedlings may appear uneven, with one plant standing upright and another leaning slightly, and it becomes tempting to treat these differences as problems that need solving rather than as part of a normal adjustment period.

This is the week where many well-meaning mistakes are made.

Overwatering often begins here, not because the soil is dry, but because watering feels like a helpful action, and doing something feels better than waiting. Fertilizers get added too early. Seeds get replanted prematurely. Plants get thinned before they have had a chance to establish themselves.

What's important to understand is that unevenness at this stage is expected.

Raised beds, despite their neat structure, still contain small variations in moisture, temperature, and airflow. Plants respond to these micro-conditions individually, not in unison, and early differences often even out on their own if given time.

During this week, observation is far more useful than intervention.

Water when the top inch of soil feels dry, not according to a schedule, and resist the urge to "help" unless there is a clear, widespread issue across the entire bed. If you are debating whether something is wrong, it usually isn't.

What People Secretly Google During This Phase

This is also the week when many beginners start searching for reassurance online, often late at night, phrasing their questions in ways that reveal just how personal the uncertainty feels.

"How long until seeds sprout?"

"What if nothing comes up?"

"Should I replant already?"

"Did I kill the seeds?"

"Why does everyone else's garden look better than mine?"

These questions are rarely about information alone. They are about fear of failure, fear of having wasted time or money, and fear of not being "good at" something that was supposed to be simple.

What most search results don't say clearly enough is this: the timeline you are on is normal.

Seeds sprout at different rates. Some take days, others take weeks. Seedlings adjust unevenly. Beds settle. Soil behaves differently after watering than it did before.

None of this means you failed.

The internet tends to show gardens at their most photogenic stages, not during the quiet, uncertain beginnings, and comparing your day-five bed to someone else's day-forty success story is a reliable way to create unnecessary discouragement.

Days Eight to Fourteen: When Things Start to Shift

Somewhere during the second week, often without a single dramatic moment, the bed begins to feel different.

More seedlings appear. Existing plants hold themselves upright for longer. Leaves look a little more purposeful. The garden starts to resemble something intentional rather than experimental.

This is when confidence usually returns, not because everything looks perfect, but because the system is clearly working.

Growth is still uneven. Some plants are ahead of others. Some seeds may never sprout at all. But the overall direction is clear.

This is also the stage where restraint continues to matter. The temptation now is to accelerate growth by adding inputs, adjusting soil, or watering more frequently, but most of the time, staying the course produces better results than pushing for speed.

What to Look For When You Check the Bed

By this point, it helps to know what is actually worth noticing when you check on the garden, because without that guidance, everything can start to feel equally important.

Look for general posture rather than perfection. Are plants upright more often than not. Are leaves oriented toward the light. Does the soil hold moisture without staying soggy.

Feel the soil a little below the surface rather than judging by appearance alone, since surface dryness does not always mean the bed needs water.

Notice patterns rather than individual outliers. One struggling plant is information. An entire bed struggling is a signal.

And then, just as importantly, leave without changing anything unless there is a clear reason to do so.

It also helps to know what counts as a genuine early red flag, because without that context, everything can feel like a red flag, especially if you've had plants die on you in the past and you're trying to prevent a repeat.

A real red flag is when the soil stays wet for days and starts to smell sour, because that usually means the roots are sitting in conditions they can't breathe in. Another real red flag is when seedlings collapse at the base, as if someone pinched the stem near the soil line, because that can be damping off, and once you see it, you want to remove affected seedlings and let the surface dry slightly between waterings.

But most of what beginners worry about in the first two weeks is not that.

A seedling leaning slightly is not a crisis. A leaf with a small blemish is not a crisis. One patch sprouting before another patch is not a crisis. Even a plant that looks a bit sulky after transplanting is usually just adjusting, especially if mornings look better than afternoons and the plant perks up again once the heat passes.

If you want a calm way to read what you're seeing, ask yourself a simpler question: does the bed look worse every day, or does it look different every day. Worse means a pattern of decline. Different usually means growth and adjustment.

And if you're still unsure, give it forty-eight hours before you do anything big. Forty-eight hours is long enough for a pattern to appear, but short enough that you don't feel like you're ignoring something important, and it prevents the most common beginner mistake, which is making a major change based on a single moment of worry.

Early "Responsible" Actions That Backfire

Many early mistakes come from trying to be diligent rather than careless.

Watering on a strict schedule instead of by feel often leads to oversaturated soil. Replanting seeds too early interrupts germination that was already underway. Adding fertilizer "just to help" can overwhelm young plants that are not ready to process it yet.

These actions feel responsible because they involve effort, but effort is not always what plants need most in the early stages.

Often, what they need is time.

This is where it helps to separate care from control.

Care is consistent watering based on what the soil actually feels like, not what the calendar says, and noticing whether the bed is drying out faster on a windy day or holding moisture longer after an overcast stretch, then adjusting gently.

Control is changing three things at once because you feel nervous.

Control looks like watering more often and then adding fertilizer and then replanting and then moving pots around, and then wondering why you can't tell what worked, because now everything is tangled together.

If you've ever had a moment where you thought, I'll just do this one extra thing to make sure it works, that's the control impulse trying to masquerade as responsibility. It comes from a good place, but it usually creates more stress than success.

A practical way to keep yourself from overcorrecting is to use a one-change rule. If you decide something truly needs adjusting, change one thing, then wait a few days and observe. If you water differently, don't fertilize the same day. If you thin seedlings, don't also replant. If you add mulch, don't also change your watering schedule immediately.

The garden needs time to respond, and you need time to learn what caused what.

This is also where beginners get tripped up by other people's confidence. Someone online will say they "feed weekly" or "boost with fish emulsion every few days," and it sounds like the reason their garden looks so good is because they're constantly doing things to it, when in reality the reason their garden looks good is usually consistency, light, decent soil, and time. The extra inputs are just their preferred routine, not the magic key.

If you want the low-effort version that works for most people, keep the soil evenly moist, add mulch once plants are established enough not to be buried by it, and hold off on fertilizer until you can see real growth, not just survival.

By the end of the second week, you might notice a strange shift, where you start feeling slightly bored with the bed, and that's actually a good sign, because boredom usually means the anxiety is dropping. You're no longer hunting for evidence that you failed. You're simply letting the garden exist, which is exactly the mindset that allows it to turn into something sustainable instead of another project you have to manage.

Learning to Observe Without Interfering

One of the most valuable skills you begin developing during these first two weeks is the ability to observe without immediately acting.

This skill does not come naturally to most people, especially those who are used to fixing problems quickly, but it becomes easier with practice, and it pays off later when genuine issues arise and thoughtful responses matter.

Gardening rewards calm attention far more than constant action.

When It's Okay to Step Away

Life continues whether the garden is ready or not.

Missing a day or two of checking rarely causes harm in a raised bed with decent soil and consistent moisture, and allowing yourself to step away without guilt prevents burnout, which is one of the most common reasons beginner gardens get abandoned.

Plants are more resilient than they appear. They do not require perfection to thrive.

The Moment You Realize You're Not Panicking Anymore

At some point, often without noticing exactly when it happened, you realize that you're no longer scanning the bed for problems every time you look at it.

You notice growth without searching for it. You trust the soil to hold moisture. You check in, then move on with your day.

That quiet shift is confidence.

It does not arrive with fireworks. It arrives with calm.

What to Do Next

As the second week comes to a close, your focus can begin to shift toward spacing, airflow, and light, topics we'll cover in the next chapter.

For now, remember this.

If the soil is moist but not soggy, if plants are upright more often than not, and if something new appears every few days, you are doing enough.

The first two weeks are not about proving anything. They are about learning the rhythm of the garden, and once you've learned that, everything that follows becomes easier.

PART IV:

Keep It Alive With Minimal Effort

Chapter 9:

The Low-Effort Watering Routine

If there is one thing that quietly ruins more beginner gardens than anything else, it is watering.

Not because people don't care. But because they care *too much*, in the wrong way.

Watering feels like the most obvious responsibility. Plants need water, so when something looks off, the instinct is to water more, water differently, water on a schedule, water with precision. That instinct is understandable, but it's also where many problems begin.

This chapter is not about mastering watering. It's about **de-escalating it**. Most watering mistakes happen because people try to be precise too early. Plants do not need perfect schedules. They need consistency within a reasonable range.

When watering is simple, consistent, and a little boring, raised beds thrive. When watering becomes a constant adjustment project, things unravel.

The goal here is to give you a routine that works in the background of your life, not one that demands attention or perfection.

Why Watering Feels Harder Than It Is

Watering problems rarely come from lack of effort. They come from uncertainty.

People worry about a few things all at once. They worry if they water too much, or too little, they worry if the soil is drying out underneath, if the surface is misleading, if today is hotter than yesterday…

When all of those questions are spinning at the same time, watering stops feeling simple and starts feeling like guesswork.

Raised beds are actually easier to water than in-ground gardens, because drainage is better and soil structure is more predictable. But that advantage only shows up when you stop trying to micromanage it.

Plants are far more tolerant than we give them credit for. What they dislike most is constant fluctuation. Occasional inconsistency will not ruin a healthy plant.

What Plants Are Really Asking For

Plants don't need a precise amount of water. They need **access** to water over time.

Roots grow toward moisture. They adapt to patterns. When watering is consistent, roots grow deeper and stronger. When watering is erratic, roots stay shallow and plants become fragile.

This is why frequent, shallow watering often causes more trouble than slightly less frequent, deeper watering. The surface looks moist, but roots never learn to explore.

In a raised bed, your job is not to keep the soil wet.

Your job is to keep it **evenly moist at root level**.

That distinction alone solves a lot of confusion.

The Biggest Watering Myth Beginners Believe

Many people believe that raised beds dry out constantly and therefore need daily watering.

Sometimes they do. Often they don't.

Raised beds drain well, which is good, but good drainage does not mean instant dryness. Soil that is well mixed and structured holds moisture surprisingly well beneath the surface.

The mistake is judging moisture by appearance alone. If you are unsure whether to water, wait and check the soil first.

Dry soil on top does not automatically mean dry soil below. In fact, the surface often dries first while deeper layers remain perfectly usable.

If you water every time the top inch looks dry, you may be watering far more than the plants need.

A Better Way to Check Moisture

Instead of looking at the surface, use touch.

Push your finger a few inches into the soil near a plant. If the soil feels cool and slightly damp, there is moisture where roots are working. If it feels dry and crumbly deeper down, it's time to water. If the top few centimeters of soil are dry but it is still moist below, do not water yet.

This method is simple, reliable, and free.

It also has a calming side effect. It replaces anxious guessing with direct information.

You don't need tools for this. You need consistency.

How Often to Water, Realistically

There is no universal watering schedule. That's not an evasion. It's reality.

Watering frequency depends on temperature, wind, plant size, soil mix, and how established the plants are. That's why rigid schedules fail so often.

What works better is a **pattern**, not a calendar.

In general terms, newly planted seeds and seedlings need more frequent attention because their roots are shallow. As plants establish, watering can become less frequent but more thorough.

In many situations, raised beds settle into a rhythm where watering every few days is enough, especially once plants are established. In hotter weather, that interval shortens. In cooler or cloudy stretches, it lengthens.

Instead of asking "How often should I water," ask "Is the soil at root level still moist."

That question adapts automatically to conditions.

Deep Watering Without Making It Complicated

Deep watering sounds technical, but it's not.

It simply means watering slowly enough that moisture penetrates beyond the surface. In a raised bed, this usually means watering until the soil is evenly moist several inches down.

This does not require special equipment. It requires patience.

Watering too quickly causes runoff or uneven absorption. Watering gently allows the soil to take in moisture properly.

Once the bed is watered deeply, you can usually wait longer before watering again. That gap is where roots strengthen.

Overwatering Looks Like Something Else

One reason overwatering is so common is that it often looks like underwatering.

Leaves may yellow. Growth may slow. Plants may look limp.

The instinct is to water more.

In reality, roots may be struggling because the soil is staying too wet and oxygen is limited. Adding more water worsens the problem.

This is where raised beds help, because excess water can drain. But if watering is constant, even raised beds can become waterlogged.

If plants look unhappy and the soil feels wet below the surface, pause. Give the bed time to dry slightly. Observe before reacting. Doing nothing for a day is sometimes the most helpful response.

Watering less is sometimes the fix.

Underwatering Is Usually Obvious

Underwatering tends to announce itself clearly.

Leaves wilt, often in the heat of the day. Soil feels dry several inches down. Plants may perk up temporarily after watering and then droop again if moisture does not reach the roots.

This is easier to diagnose than overwatering, and easier to correct.

A thorough watering that reaches root level usually resolves the issue quickly.

The key is not to panic and water lightly over and over. That keeps roots shallow and prolongs the problem.

Morning vs Evening Watering

You may hear strong opinions about watering at certain times of day.

The truth is calmer than the advice suggests.

Watering in the morning is often recommended because it allows leaves to dry during the day, reducing disease risk. This is a good habit when possible.

Watering in the evening is not automatically harmful, especially in raised beds with good airflow and drainage. In hot weather, evening watering can reduce stress.

The real issue is not timing. It's consistency and soil moisture.

If morning watering fits your routine, use it. If evening watering is more realistic, that's fine too. What matters is that watering happens regularly enough and deeply enough.

How Weather Changes Everything

Heat, wind, and sun intensity affect how quickly water moves through the bed.

Hot, windy days pull moisture from soil faster. Cooler, overcast days slow evaporation.

This is why fixed schedules struggle. Conditions change, but schedules don't.

Instead of adjusting your watering rules constantly, adjust your awareness. Check the soil more often during extreme weather. Relax during mild stretches.

Plants respond to trends, not single days.

Mulch: The Quiet Watering Helper

Mulch is one of the easiest ways to reduce watering effort, and it is often overlooked.

A layer of mulch on top of the soil helps retain moisture, moderate temperature, and reduce surface evaporation. It also suppresses weeds, which compete for water.

Mulch does not need to be fancy. Straw, shredded leaves, or untreated wood chips can all work. The key is coverage, not aesthetics.

Adding mulch can noticeably extend the time between waterings. It also makes moisture levels more stable, which plants appreciate.

This is one of those low-effort improvements that pays off immediately.

Simple Watering Setups That Actually Help

You do not need an elaborate irrigation system to water a raised bed well.

Many people do just fine with a watering can or a hose with a gentle setting. These tools give you direct feedback and encourage observation.

If you want to automate a little, simple drip lines or soaker hoses can reduce effort without removing awareness entirely. Timers can help, but they should be adjusted occasionally as conditions change.

Be cautious with fully automated setups early on. They are convenient, but they can hide problems if something goes wrong.

It's often better to understand your bed's needs first, then automate gently.

A Common Beginner Trap

Here is a pattern that shows up often.

A bed is watered consistently for the first few weeks. Plants grow well. Confidence builds. Then life gets busy. Watering becomes irregular. Plants struggle. Panic sets in. Watering becomes excessive.

The bed never settles.

Consistency matters more than intensity. A simple routine done imperfectly but regularly outperforms bursts of effort followed by neglect.

This is why it helps to tie watering to an existing habit. Morning coffee. Evening wind-down. Whatever fits naturally.

Gardens thrive when they're woven into life, not added on top of it.

When Not to Water

No Overthinking Rule #5

When in doubt, **wait one day** before watering unless the soil is dry at root level.

This is just as important as knowing when to water.

Do not water because you feel guilty. Do not water because it's on the schedule. Do not water because the surface looks dry.

Water because the **soil at root level needs it.**

Skipping a watering when soil is still moist is not neglect. It's good judgment.

Letting Plants Teach You

Plants communicate through subtle signals. Growth rate, leaf color, posture, and resilience all reflect how well water needs are being met.

You do not need to decode every signal. You just need to notice trends.

If plants look generally healthy and are growing steadily, your watering routine is working. You do not need to change it just because conditions vary slightly day to day.

Gardening improves when you stop chasing perfection and start trusting patterns.

Reality Check: Trying Harder Usually Makes Watering Worse

This deserves to be said plainly.

Most watering problems come from trying too hard. Watering improves when you intervene less and observe more.

Too much attention. Too many adjustments. Too much reaction to small changes.

Plants prefer calm conditions. They adapt well when those conditions are stable.

Your raised bed does not need constant correction. It needs steady care.

What to Do Next

Once you have a basic watering rhythm, resist the urge to keep refining it.

Stick with what works. Observe occasionally. Adjust only when something clearly changes.

In the next chapter, we'll talk about feeding plants. You'll see why fertilizing is often overdone, when it actually helps, and how to keep it simple without shortchanging your harvest.

For now, let watering become routine.

When watering feels boring, you're probably doing it right.

Chapter 10:

Feeding Plants Without Overdoing It

Feeding plants is where many well-intentioned gardeners start working against themselves.

It usually begins innocently enough. Growth slows a little. Leaves look slightly lighter than before. Someone online says vegetables are "heavy feeders," and suddenly it feels irresponsible not to add something. Fertilizer becomes a reflex rather than a response.

This chapter is here to reset that instinct. Overfeeding causes more problems for beginners than underfeeding.

Feeding is not a rescue tool. It's a **maintenance tool**, and it works best when used lightly, deliberately, and later than most beginners expect. If a plant looks healthy and is growing, feeding can wait.

If watering is about consistency, feeding is about restraint.

What Feeding Is Actually For

Before adding anything, pause and look at overall growth rather than individual leaves.

Plants need nutrients, but they don't need constant stimulation.

In a raised bed with decent soil, nutrients are already present and releasing slowly. Compost breaks down. Organic matter feeds soil life. Roots explore and access what they need over time. Early growth depends far more on root establishment than on additional inputs.

Feeding becomes useful when plants are actively growing and drawing heavily from the soil. It supports continued production. It does not create it from nothing.

That distinction matters.

Fertilizer helps plants that are already doing reasonably well do better for longer. It does not fix stress, poor placement, or inconsistent watering.

Why Early Feeding Backfires So Often

Feeding too early is one of the most common beginner mistakes, and it usually comes from impatience rather than neglect.

During the first weeks after planting, roots are still establishing themselves. Plants are adjusting to new soil, new light, and new moisture patterns. Pushing nutrients at this stage often leads to soft, rapid leaf growth without strong root support.

That kind of growth looks impressive briefly, then creates problems later. Plants flop more easily. They become more sensitive to heat. They attract more pest attention.

In other words, feeding early feels helpful but often creates fragility.

Let plants settle first. Strong roots make feeding effective later.

The Rule That Prevents Most Feeding Mistakes

Here is the simplest rule you need.

If plants are green, growing steadily, and producing new leaves, you do not need to feed them yet.

This rule sounds almost too passive, but it works because it respects how plants actually grow. Feeding should respond to sustained demand, not momentary concern.

If you feel the urge to feed "just in case," that's usually a sign to wait another week. Feeding too early is easier to undo by waiting than by trying to correct excess nutrients later.

Feeding Is Not the Same as Fixing

This is where feeding and troubleshooting often get confused.

If a plant looks stressed, pale, or stalled, feeding is rarely the first or best response. Stress usually comes from water, temperature, crowding, or recent disturbance. Adding nutrients on top of stress rarely helps and sometimes makes things worse.

Feeding supports **healthy growth**. It does not correct underlying instability.

That's why feeding belongs after planting and watering have settled into a rhythm, not before.

When Feeding Actually Helps

Feeding starts to make sense once plants are established and actively growing.

This often happens later in the season, or after repeated harvesting. Leafy greens that are cut regularly, for example, pull nutrients from the soil faster than soil can replenish them on its own.

You may notice that regrowth slows slightly, or that leaves are smaller than earlier harvests. This is not a crisis. It's information.

A modest feed at this point replenishes what's been used and supports continued production without forcing growth.

Timing is what makes feeding helpful rather than disruptive.

The Simplest Feeding Options That Work Well

You do not need variety here. You need reliability.

Compost added to the soil surface is one of the gentlest and most effective feeding methods available. It improves soil structure, feeds soil organisms, and releases nutrients gradually. It is forgiving and hard to overdo.

A balanced, all-purpose fertilizer labeled for vegetables is another straightforward option. Used at the lower end of the recommended rate, it supports growth without overwhelming plants.

Liquid feeds act faster and can be useful when plants are clearly established and producing heavily, but they also move through the soil more quickly. That makes them easier to overapply if you're not careful.

Whichever option you choose, consistency and moderation matter more than strength.

Organic vs Synthetic, Without the Debate

You don't need to pick a side to grow vegetables.

Organic fertilizers tend to release nutrients slowly and build soil health over time. They are generally more forgiving, which makes them appealing for beginners.

Synthetic fertilizers deliver nutrients quickly and predictably. Used carefully, they can be effective. Used aggressively, they can stress plants and wash through soil.

The practical difference is not ideology. It's **speed and margin for error**.

If you prefer a wider safety margin, choose gentler options. If you're comfortable following directions carefully, faster options can work too.

Neither choice replaces observation.

How Often to Feed, Without Turning It Into a Schedule

Feeding does not need to be frequent.

In many raised beds, feeding once mid-season and again only if production continues heavily is enough. Some beds thrive with nothing more than compost added between plantings.

The idea that vegetables need constant feeding comes from high-intensity production systems, not home gardens designed for simplicity.

Let the plants lead. If growth remains steady and leaves stay healthy, feeding can wait.

Feeding Works Best With Stable Watering

Nutrients move with water. That means feeding and watering are always linked.

If watering is inconsistent, feeding becomes unpredictable. Nutrients may concentrate in dry soil or wash through when water suddenly increases.

This is why it's important to establish a watering rhythm before feeding regularly. Stable moisture allows roots to access nutrients evenly.

If feeding doesn't seem to help, look at watering first rather than adding more fertilizer.

What Feeding Should Feel Like

Feeding should feel uneventful.

You apply something modest. Growth continues. Nothing dramatic happens.

If feeding creates a sudden surge, something is off. Either timing, quantity, or both.

Healthy feeding supports momentum rather than creating it.

Common Feeding Missteps, Reframed

Feeding multiple products at once often comes from fear, not need. Compost plus slow-release fertilizer plus liquid feed is rarely necessary and often counterproductive. Use one feeding method at a time, not several.

Responding to every small visual change with feeding turns maintenance into micromanagement.

Trying to "make up for" weak growth with nutrients usually masks the real issue instead of solving it.

These missteps are common, and they're easy to avoid once feeding stops being emotional.

Letting Soil Carry More of the Load

One of the long-term benefits of raised beds is that soil improves with time.

Organic matter breaks down. Microbial life increases. Nutrients cycle more efficiently. Over time, the need for feeding often decreases rather than increases.

Adding compost periodically and avoiding overdisturbance supports this process quietly in the background.

Feeding becomes a supplement, not a dependency.

When Doing Nothing Is Still the Right Choice

It bears repeating, because this is where confidence really develops.

If plants are healthy, productive, and stable, doing nothing is not neglect. It is good judgment.

Gardening does not reward constant intervention. It rewards knowing when to step back. Restraint is a skill, and it develops faster than people expect.

What to Do Next

At this point, feeding should feel demystified rather than urgent.

You know when it helps. You know when it doesn't. You know that less usually works better.

In the next chapter, we'll move from prevention to clarity. You'll learn how to quickly diagnose issues when something truly does change, without spiraling or stacking fixes.

For now, let feeding remain what it should be.

A quiet support act, not the main event.

PART V:

When Something Goes Wrong

Chapter 11:

Let's Troubleshoot

At some point, something in your raised bed will look off.

A leaf turns yellow. Growth slows down. A plant that looked fine last week suddenly seems tired, smaller, or less upright than its neighbors. This is the moment when many people either panic or quietly disengage, and neither response helps, because most garden problems are not emergencies, even though they often feel that way in the moment. In most cases, time is on your side.

This chapter exists to slow that moment down.

Not to dismiss your concern, and not to tell you to relax, but to give you a simple way to assess what you are seeing without spiraling into fixes, products, or drastic decisions. Most garden issues are signals, not failures, and almost all of them can be narrowed down quickly if you ask the right questions in the right order.

You do not need to know everything about plants. You need a calm method for deciding what matters and what does not. This method exists to protect you from reacting too quickly.

Why Problems Feel Bigger Than They Are

Garden problems rarely arrive dramatically. They creep in quietly. A leaf looks slightly paler than it did last week. A plant seems a bit shorter than the one next to it. A patch of soil stays damp longer than you expected. Because the change is subtle, your brain fills in the gaps, and those gaps tend to fill with doubt rather than information.

You start wondering whether the soil is wrong, whether the weather ruined everything, whether you planted at the wrong time, whether you should add fertilizer, water more, pull the plant out, or replace it with something easier. The situation feels urgent even when nothing is actually escalating.

This mental spiral is far more damaging than the original issue.

Raised beds, by design, reduce complexity. You chose the location. You chose the soil. You know roughly how you have been watering. That dramatically narrows the field of possible problems, even if it does not feel that way when you are standing in front of a plant that looks unhappy.

Most beginner problems fall into a handful of categories. Once you identify which category you are dealing with, the solution is usually simple, or surprisingly often, no action at all.

The Only Question That Matters First

Before you look closely at any leaf, ask yourself one question.

Is the problem affecting one plant or many?

This distinction saves an enormous amount of time and stress.

If one plant looks off while the others nearby look fine, the issue is usually local. A weak seedling, transplant shock, a damaged root, or a single pest. These problems rarely require bed-wide intervention.

If several plants show the same symptoms at the same time, the issue is usually systemic. Watering, soil moisture, temperature, light, or feeding patterns are the likely causes.

Do not skip this step. This single question prevents more unnecessary fixes than any other diagnostic step. Treating a whole bed for a problem that affects one plant is one of the fastest ways to create new problems that did not exist before.

STEP ONE: Check Water Before Anything Else

Watering causes more beginner problems than any other factor, and it is also the easiest thing to misinterpret.

Before you assume nutrient deficiency, disease, or pests, check the soil itself.

Push your finger a few inches into the soil near the affected plant. Not just the surface, and not right against the stem. A few inches down is where the truth usually is.

If the soil feels wet and heavy, lack of water is not the problem. If it feels dry at root level, underwatering is very likely involved.

Many symptoms overlap. Yellowing leaves, slow growth, and drooping can all result from either too much or too little water. The soil tells you which one you are dealing with far more reliably than the plant's appearance alone.

If the soil is too wet, the fix is almost always restraint. Acting immediately usually makes wet soil problems worse. Let the bed dry slightly. Do not add fertilizer. Do not water again to flush things out. Time, airflow, and patience usually solve the issue.

If the soil is too dry, water deeply and evenly. One thorough watering that reaches root depth is more effective than several light ones that only wet the surface.

If you remember only one diagnostic step, remember this one.

Why Overwatering Is So Common

Overwatering rarely comes from neglect. It comes from concern.

When something looks wrong, watering feels like a caring action, and because it is visible and immediate, it creates the sense that you are helping, even when the soil does not need more moisture. In raised beds, especially those with good drainage, it is easy to assume that more water cannot hurt, but roots need air as much as they need moisture, and saturated soil removes that air.

This is why many plants look worse after being watered again, not better.

Understanding this shifts watering from a reflex into a decision, which alone prevents a large percentage of early failures.

STEP TWO: Look at the Pattern, Not the Color

Leaf color gets far more attention than it deserves.

Instead of focusing on the exact shade of yellow or green, look at the pattern of change.

Are older leaves affected first, or is new growth showing issues. Is discoloration spreading rapidly, or staying contained. Is the plant still producing new leaves, even if they are smaller than expected.

Older leaves yellowing while new growth remains healthy is often normal, especially as plants mature or redirect energy. New growth showing problems can indicate stress, but stress does not automatically mean deficiency or disease.

A plant that is still growing, even slowly, is usually coping. Coping plants rarely need immediate intervention. A plant that has completely stopped growing is the one asking for attention.

A Missing Piece: Timing Matters More Than Symptoms

One of the most overlooked diagnostic clues is timing.

Ask yourself when the change appeared. Was it immediately after planting. After a sudden heatwave. After several days of rain. After a missed watering. After thinning or harvesting.

Plants react to changes with a delay. A leaf turning yellow today often reflects something that happened several days earlier. If you treat today's symptom without considering yesterday's conditions, you often fix the wrong thing.

This is why drastic changes made in response to a single bad day rarely help. The garden is always responding to a short history, not just the present moment.

Learning to link symptoms to recent events builds confidence faster than memorizing deficiency charts ever will.

The Most Common Problems, Explained Without Drama

A small number of issues account for the majority of beginner concerns.

Transplant shock is extremely common. Seedlings often pause after planting. Leaves may droop slightly. Growth may stall for a week or two. This is normal behavior as roots establish themselves in new soil.

Inconsistent watering is another frequent cause of slow growth. Not extreme drought or flooding, but fluctuation. Plants prefer rhythm. When moisture swings back and forth, growth slows even if nothing else is wrong.

Nutrient depletion usually shows up later in the season, especially in beds with frequent harvesting. Leaves may become paler overall and growth may slow evenly across the plant. This is when gentle feeding or compost top-ups help.

Pest damage tends to look irregular. Holes, chewed edges, or visible insects. If damage is minor and growth continues, intervention is often unnecessary.

A Calm Way to Rule Out Pests

Pests are blamed for far more problems than they actually cause.

Before treating anything, look closely. Turn leaves over. Check stems. Look for insects, eggs, or clear bite patterns.

If you do not see evidence, pests are unlikely to be the issue.

Even when pests are present, raised beds tolerate low levels surprisingly well. Plants are resilient. A few holes do not mean the crop is lost.

Action makes sense only when damage is increasing rapidly or affecting many plants. Otherwise, observation is often enough.

When Feeding Is the Wrong Fix

Fertilizer is often treated as a cure-all, but early feeding frequently creates more problems than it solves.

If watering is inconsistent, feeding will not fix it. If roots are stressed, extra nutrients can overwhelm them. If soil is already rich, feeding creates imbalance rather than improvement.

Feeding helps when plants are actively growing and have used up available nutrients. It does not help when plants are stressed for other reasons.

When in doubt, wait. Observation almost always provides better guidance than early intervention.

Another Missing Skill: Learning What Not to Touch

Many beginner mistakes come from touching too much.

Straightening a slightly leaning plant. Firming soil repeatedly around stems. Repositioning seedlings that are simply adjusting. These actions feel helpful, but they interrupt root establishment and increase stress.

Plants anchor themselves through movement and resistance. A bit of wobble strengthens stems. Slight unevenness encourages roots to spread. Interfering constantly removes the very signals plants use to adapt.

Doing less often produces stronger plants. Restraint is not inaction, but informed choice.

A Simple Five-Minute Diagnostic That Actually Works

The diagnostic process is intentionally simple.

First, determine whether the issue affects one plant or many.

Second, check soil moisture at root depth.

Third, look for visible pest damage.

Fourth, consider timing and recent changes.

Fifth, decide whether the problem is worsening or stable.

No Overthinking Rule #6

Change **one thing at a time**, or you will never know what worked.

If the situation is stable or improving, doing nothing is often the correct choice.

If it is clearly worsening and you have identified a likely cause, address that one thing only. Multiple changes at once make it harder to know what actually helped.

Problems You Can Usually Ignore

This surprises many people.

A single yellow leaf on an otherwise healthy plant is rarely a problem.

Slow growth during cooler weather is normal.

Uneven seedling size happens.

Minor cosmetic damage does not affect harvest.

Plants are living systems, not manufactured products. They grow unevenly, respond individually, and adapt constantly.

Chasing visual perfection often creates real problems where none existed.

When Pulling a Plant Is the Right Decision

Sometimes the best choice is to remove a plant.

If one plant is clearly failing while others thrive, it may simply be weak. Removing it frees space, light, and nutrients for healthier plants.

This is not giving up. It is pruning.

Raised beds make this easier because space is limited and competition matters. Letting go of one plant often improves the entire bed.

A Walk-Through of a Real Scenario

You notice several leaves on a lettuce plant turning yellow. The plant is still growing, but more slowly than last week.

You check the soil and find it consistently damp. You have been watering daily because the weather has been warm.

This points to overwatering, not nutrient deficiency.

The fix is simple. Water less often. Allow the soil to dry slightly between waterings. Do not fertilize. Do not replant.

Within a week, growth usually improves.

This is how most garden problems resolve. Quietly, once pressure is removed.

Why Raised Beds Recover So Well

Raised beds recover quickly because drainage improves easily, soil structure responds to small changes, and roots are not fighting compacted ground.

Mistakes do not linger as long as they do in in-ground gardens.

This means you can afford patience.

Learning Without Keeping Score

It is tempting to treat every issue as a lesson you must record.

You do not need to.

Gardening knowledge accumulates naturally. You notice patterns. You adjust instinctively. Confidence builds quietly.

The goal is not mastery. It is steady improvement with less effort.

What to Do Next

The next time something looks wrong, pause.

Run the five-minute diagnostic. Check water. Look for patterns. Observe for a few days.

Most issues resolve with small adjustments or time.

In the next chapter, we will focus specifically on pests and weeds, and how to handle them without turning your garden into a battleground.

For now, remember this.

A calm response solves more garden problems than any product ever will. Most successful gardeners are not the most active ones, they are the most patient.

Chapter 12:

Pests and Weeds Without Turning It Into a War

This is the chapter many people brace themselves for long before there is any real reason to, because pests and weeds carry a kind of emotional weight that far exceeds the damage they usually cause, especially in raised beds that are already doing a lot of the containment and prevention work for you.

Somewhere along the line, gardening picked up the idea that insects are enemies, weeds are failures, and a good garden is one that looks untouched by nature. That idea is not only unrealistic, it actively makes gardening more stressful than it needs to be.

The calmer and far more useful truth is this.

Most gardens have pests. Most gardens have weeds. Most successful gardeners learn how to **live with a low level of both** without turning every appearance into a problem that needs solving.

This chapter is about reaching that mindset quickly, so you don't waste energy fighting battles that don't matter.

The First Question That Changes Everything

Before identifying insects, researching treatments, or assuming something has gone wrong, there is one question that should always come first.

Is the plant still growing.

If growth is continuing, even slowly, the situation is almost always manageable. Plants can lose leaves, tolerate chewing, and share space with other organisms while still producing food. Growth is the signal that the system is functioning.

If growth has stopped completely and damage is increasing rapidly, that's different. That's when attention is warranted.

This single check filters out the majority of unnecessary interventions.

Why Raised Beds Are Already on Your Side

One of the quiet advantages of raised beds is containment.

You're not dealing with an open ecosystem stretching across a yard or field. You're working within defined borders, controlled soil, and plants that are easy to see and reach. That alone reduces pest pressure and makes weeds easier to manage.

Raised beds also tend to support healthier plants earlier, and healthy plants are far less attractive to pests than stressed ones. Many insect problems are secondary issues that appear when plants are already struggling for other reasons.

In other words, if you've followed the earlier steps in this book, you've already done most of the preventative work without realizing it.

Most Insects Are Background Noise

It's important to say this plainly, because it changes how you react.

Seeing insects does not mean you have a pest problem.

Gardens are living spaces. Insects move through them constantly. Many insects do nothing at all to your plants. Some feed on other insects. Others are passing through without any interest in staying.

The presence of insects is not the same as damage, and damage is not the same as danger.

Problems arise when populations explode quickly and plants can't keep up. That's a scale issue, not a presence issue.

What Pest Damage Usually Looks Like in Real Life

In beginner raised beds, pest damage is usually cosmetic at first.

Small holes in leaves.

Edges that look nibbled.

A leaf that's missing a corner.

This kind of damage often looks alarming when you first notice it, but it rarely affects harvest in any meaningful way. Plants have more leaf surface than they strictly need, and they are very good at compensating.

More serious pest issues tend to show a pattern over time rather than appearing suddenly. Damage increases daily. New leaves emerge already compromised. Growth slows noticeably.

That progression is what matters, not a single imperfect leaf.

How to Look Without Obsessing

You do not need to inspect every plant every day.

A quick glance while watering or harvesting is enough. Occasionally turning a leaf over and looking underneath gives you useful information without turning the garden into something you manage with suspicion.

If you see nothing obvious and damage isn't escalating, you can stop looking.

Constant inspection doesn't prevent problems. It just creates anxiety.

When Intervention Actually Makes Sense

Intervention is appropriate when three things are true at the same time.

Damage is increasing quickly.

Growth is clearly affected.

You can identify a likely cause rather than guessing.

When those conditions are met, start with the least disruptive option.

Physical removal is often enough. Knocking insects off with water, removing affected leaves, or manually reducing numbers can dramatically slow a problem without introducing new variables.

Barriers are another effective option, especially for young plants. Protecting seedlings during their most vulnerable stage often prevents the problem entirely, because many pests target plants only when they are small.

Chemical treatments, including organic ones, should be the last step, not the first. They affect more than just the target insect, and once used, they often need to be used repeatedly.

The goal is control, not eradication.

Why Overreacting Usually Backfires

One of the hardest lessons to learn is that aggressive responses often make pest problems worse over time.

Broad treatments remove beneficial insects along with pests, which creates a vacuum that pests are very good at filling. The next wave often arrives faster and stronger than the first.

Raised beds do best when the system is allowed to stabilize. Small disruptions are fine. Constant resets are not.

If a plant can tolerate some damage while continuing to grow, letting it do so is often the most effective long-term strategy.

Weeds Are a Timing Problem, Not a Moral One

Weeds trigger a different kind of reaction, often tied to neatness and control rather than fear of loss.

Here's the grounding reality.

Weeds grow because seeds exist. They arrive through wind, compost, soil, birds, and shoes. Their presence does not say anything about your effort or competence.

In raised beds, weeds are usually fewer and easier to remove than in-ground gardens. That's already a win.

The only thing that matters with weeds is timing.

The Only Weed Rule You Actually Need

Pull weeds while they are small.

That's it.

Small weeds come out easily, disturb very little soil, and haven't yet competed meaningfully with your plants. Large weeds are annoying because they were ignored while they were small.

A few minutes here and there prevents hours of work later.

This approach keeps weed management proportional rather than overwhelming.

Why Mulch Does Most of the Work for You

Mulch is one of the simplest, most effective tools you have, and it works quietly in the background.

By covering the soil surface, mulch blocks light that weed seeds need to germinate. It also keeps moisture levels more stable and reduces temperature swings, which benefits your plants at the same time.

Mulch does not need to be thick or fancy. Coverage matters more than perfection. Refresh it when it thins out. Let it do its job.

When mulch is in place, weed pressure drops dramatically, and so does the mental load of managing it.

When Weeds Are Actually Useful Information

Occasionally, weeds point to something worth adjusting.

Bare soil invites weeds, which is a signal to mulch or plant more densely.
Repeated disturbance brings up more seeds, which is a signal to disturb the soil less.

But most of the time, weeds are simply opportunistic, not diagnostic.

Remove them calmly and move on.

Knowing When to Ignore Both Pests and Weeds

This is where confidence really shows up.

If plants are growing, producing, and generally healthy, minor pest damage and occasional weeds are background noise. They do not require correction.

Ignoring manageable issues frees time and attention for harvesting, which is the actual point of the garden.

A garden does not need to look controlled to be productive.

A Reality Check About Control

You are not running a clean room. You are tending a small ecosystem.

Complete control is not possible, and trying to achieve it leads to constant work with diminishing returns. What *is* possible is a raised bed that produces food reliably with minimal effort and a reasonable tolerance for imperfection.

That's the goal worth holding onto.

What to Do Next

As you move forward, let pests and weeds become information rather than threats.

Notice patterns. Intervene gently when needed. Ignore what doesn't matter.

In the next chapter, we'll return to something more rewarding and far less stressful: harvesting in a way that actually increases how much food your plants produce over time.

That's where the garden starts paying you back for your patience.

PART VI:

Harvest, Reset, and Keep It Simple

Chapter 13:

Harvesting for More Food

(Not Just Once)

Harvesting is the point where gardening finally feels like it's paying you back, but it's also the stage where many beginners unintentionally limit their own results, not through neglect, but through hesitation.

There is a quiet fear around harvesting that doesn't get talked about much. People worry they'll take too much. They worry they'll cut the wrong thing. They worry they'll ruin a plant that took weeks to grow by being impatient or careless at the wrong moment.

So they wait.

And while waiting feels cautious and responsible, it often does the opposite of what they intended.

Most vegetables are designed to be harvested. In fact, many of them grow *better* because you harvest them. The act of removing leaves, stems, or fruits sends a clear signal to the plant to keep producing rather than slow down or shut off.

This chapter is about harvesting in a way that feels confident, measured, and productive, without turning it into another thing to overthink.

Why Harvesting Is Part of Growing, Not the End of It

It helps to shift how you think about harvesting before talking about technique.

Harvesting is not the final step in a plant's life. It's feedback. When you harvest correctly, you are telling the plant that conditions are good and that continued growth makes sense.

Many vegetables respond to harvesting by producing more leaves, more shoots, or more fruit. When harvesting is delayed too long, plants often move into a different phase, focusing on flowering or seed production instead of edible growth.

In other words, harvesting is not something you do *to* plants. It's something you do *with* them.

Once you understand that relationship, harvesting becomes less intimidating and more intuitive.

The Biggest Harvesting Mistake Beginners Make

The most common mistake is waiting for vegetables to look "fully grown," which usually means larger than ideal.

Seed packets and plant labels often show mature plants at their biggest and most impressive. Those images are not always showing the best harvest stage. They are showing potential, not preference.

Many vegetables taste better, regrow faster, and stay productive longer when harvested earlier rather than later. Overgrown vegetables tend to become tougher, more bitter, or less reliable.

Harvesting sooner feels risky. Harvesting late feels safe. In reality, the opposite is often true.

How to Tell When Something Is Ready Without Charts

You do not need exact measurements to know when most vegetables are ready.

For leafy greens, readiness usually looks like leaves large enough to use, but still tender. If a leaf looks appetizing, it probably is. Waiting for maximum size rarely improves flavor.

For herbs, harvesting once plants have multiple healthy stems encourages bushier growth. You don't need to wait for flowering. In many cases, harvesting before flowering preserves flavor.

For fruiting vegetables, size, color, and firmness matter more than time. Most fruits are best harvested once they reach their typical shape and color, not once they've sat on the plant as long as possible.

If you're unsure, harvesting a small amount as a test is rarely a mistake.

Cut-and-Come-Again: Why This Changes Everything

Many beginner-friendly vegetables fall into a category called cut-and-come-again. This includes most leafy greens and many herbs.

Instead of pulling the entire plant, you harvest outer leaves or stems while leaving the center intact. The plant continues growing and produces new leaves over time.

This approach dramatically increases total yield from a small space. It also spreads harvests out, which feels less wasteful and more manageable.

The key is moderation. Removing some growth regularly works better than removing everything at once.

Think of it as pruning rather than harvesting, and the results improve naturally.

How Much to Harvest Without Hurting the Plant

A simple rule helps here.

Avoid removing more than about one-third of a plant at a time.

This gives the plant enough leaf area to continue photosynthesis while still benefiting from the harvest signal. It's not an exact number. It's a guideline that keeps things balanced.

If you accidentally take more, it's rarely fatal, especially with healthy plants. But using restraint consistently leads to stronger regrowth and longer production.

Harvesting Tools: Simple Is Enough

You don't need specialized tools to harvest well.

Clean hands, a pair of scissors, or basic garden snips are enough for most tasks. What matters more than the tool is cleanliness. Using clean tools reduces stress and prevents disease transfer.

Avoid tearing or crushing stems when possible. Clean cuts heal faster and reduce the chance of rot.

That said, perfection is not required. Plants are resilient, and small imperfections in technique are rarely consequential.

Harvesting as Observation Time

Harvesting is one of the best moments to notice how your plants are actually doing.

You're close enough to see leaf color, texture, new growth, and any developing issues. This quiet observation often reveals things you might miss when just watering or walking past.

If something looks off, note it. You don't need to act immediately. Awareness alone improves decision-making later.

This is another reason frequent, light harvesting works so well. It keeps you connected without making the garden feel demanding.

When Harvesting Feels Scarce

Early harvests often feel small.

You may wonder whether it's even worth picking a handful of leaves or a single herb stem. This hesitation is normal, especially if you're used to buying produce in larger quantities.

Remember that harvesting encourages more growth. Leaving everything untouched does not increase future abundance. It often reduces it.

Small, early harvests are an investment in larger ones later.

Overharvesting and How to Recover

Overharvesting happens. Sometimes enthusiasm gets ahead of judgment. Sometimes dinner plans do.

Most plants recover surprisingly well, especially if the soil is healthy and watering is consistent. Growth may slow briefly, then resume.

If you've taken too much, the best response is to pause harvesting and let the plant rebuild. Avoid feeding immediately unless the plant was already well established and producing heavily.

Time and consistency usually do more than intervention here.

Harvest Timing and Flavor

When you harvest can matter as much as how.

Many leafy vegetables taste best when harvested during cooler parts of the day, often in the morning. Leaves are more hydrated, textures are better, and flavors are cleaner.

This does not mean you must harvest at a specific hour. It means that if you have a choice, cooler conditions usually improve quality.

If you harvest at a less ideal time, the vegetables are still usable. This is a refinement, not a requirement.

Letting Some Plants Finish Their Cycle

Not every plant needs to be harvested indefinitely.

Some vegetables produce once and are done. Others slow naturally as conditions change. Letting a plant finish its cycle is not waste. It's part of the rhythm of the bed.

As plants decline, space opens up for replanting, which we'll cover in the next chapter.

Harvesting is not about squeezing every possible leaf out of a plant. It's about maintaining steady, enjoyable production.

A Common Beginner Scenario, Reframed

You grow lettuce and hesitate to cut it. Weeks pass. The leaves get larger, then bitter. The plant bolts and becomes less usable.

If you had harvested earlier and regularly, you would likely have enjoyed multiple mild, tender harvests over a longer period.

This is not a mistake so much as a learning moment. Almost everyone experiences it once.

Harvesting Builds Confidence Quietly

There is something grounding about using food you grew yourself, even in small amounts.

Each harvest reinforces that the system works. That your decisions were good enough. That effort turned into something tangible.

This confidence carries forward into the next steps, including replanting and extending the season.

What to Do Next

As harvesting becomes part of your routine, you'll start noticing gaps. Spaces where plants are slowing down. Areas where something could grow next.

In the next chapter, we'll talk about replanting and season extension in a way that doesn't require starting over or turning gardening into a year-round project.

For now, harvest with intention, not hesitation.

Taking food from the garden is not the end of growth.

It's how you invite more of it.

Chapter 14:

Season Extension and Replanting (Without Starting Over)

By the time you reach this point, something important has already happened, even if you haven't named it yet. The garden has stopped feeling fragile.

Plants have grown. You've harvested food. You've seen things succeed, stall, recover, and move on. That experience changes how the bed feels, because it's no longer a one-shot experiment that could be ruined by a single wrong move. It's a working system.

Season extension and replanting build directly on that shift.

This chapter is not about pushing your garden to its limits or squeezing productivity out of every square inch. It's about **keeping things going gently**, without tearing everything out or mentally resetting back to zero.

Why Most Gardens Stop Earlier Than They Need To

Many gardens slow down not because the season is over, but because the gardener assumes it is.

A plant finishes producing. A few beds look tired. Growth isn't as exciting as it was earlier. The garden still works, but momentum drops, and with it, attention.

That pause is understandable, but it's often unnecessary.

Raised beds, especially, are well suited to replanting because the soil stays workable, drains well, and doesn't compact the way in-ground soil can. You don't have to wait for a full reset to plant something new.

You just need to know what to leave alone and what to replace.

Understanding the Natural Rhythm of the Bed

Not all plants are meant to last the same amount of time.

Leafy greens often come and go quickly, especially in warm weather. Some herbs thrive all season. Fruiting plants tend to have a longer arc, producing steadily before slowing down.

When a plant's production drops significantly and doesn't rebound after normal care, it's usually signaling that it's nearing the end of its useful cycle.

That signal is not failure. It's an opening.

Seeing plant decline as a cue rather than a problem makes replanting feel natural instead of disruptive.

What to Pull and What to Keep

This decision doesn't need to be complicated.

If a plant is no longer producing and looks stressed despite reasonable care, removing it is often the best option. Leaving declining plants in place rarely improves them, and it blocks light and space that new plants could use.

If a plant is still producing, even modestly, and looks generally healthy, there is no urgency to remove it. Let it continue at its own pace.

You don't need uniformity. A raised bed can easily hold plants at different stages of life without issue.

Mixed timelines are normal and useful.

Replanting Without Disturbing the Whole Bed

One of the advantages of raised beds is how localized work can be.

When you remove a finished plant, you don't need to dig up the entire bed. Gently loosen the soil in that specific spot, remove old roots if they're dense, and add a small amount of compost to refresh the area.

That's enough.

There's no requirement to replace all the soil, turn everything over, or "start fresh." In fact, disturbing the entire bed too often can set you back by disrupting soil structure that has already settled into a good pattern.

Targeted replanting keeps the system stable.

Choosing What to Plant Next, Simply

This is not the moment to chase complexity.

Replanting works best when you choose crops that are reliable, familiar, or quick. Leafy greens, herbs, and fast-growing vegetables fit well here because they establish quickly and don't demand perfect timing.

If you're unsure what to plant next, return to what worked earlier. Repeating success is a perfectly valid strategy.

This phase is about continuity, not experimentation.

Temperature Matters More Than the Calendar

Instead of thinking in terms of months or seasons, pay attention to conditions.

Plants respond to temperature and daylight more than dates. If daytime temperatures are still moderate and nights aren't extreme, many crops will continue growing happily.

When temperatures shift beyond a plant's comfort range, growth slows regardless of when it was planted.

This is why rigid planting calendars often create more confusion than clarity. Your bed will tell you more than a chart if you pay attention to how plants are responding.

Simple Ways to Extend the Season Without Extra Work

Season extension doesn't require structures, heaters, or special equipment to be effective.

Sometimes it's as simple as continuing to harvest, watering consistently, and letting plants take advantage of mild conditions longer than expected.

In cooler periods, placing the bed in a sheltered location, reducing wind exposure, or using mulch to moderate soil temperature can extend productivity without adding tasks.

The goal is not to force growth. It's to avoid unnecessary decline.

When Slowing Down Is the Right Choice

There will be times when replanting doesn't make sense, and that's fine.

If conditions are shifting quickly or your attention is limited, allowing the bed to wind down naturally can be the most practical decision. Gardening does not have to be continuous to be successful.

Pausing does not erase what you've built. The soil remains improved. The experience carries forward.

Knowing when to stop is as valuable as knowing when to continue.

A Common Replanting Hesitation, Reframed

Many people hesitate to replant because it feels like committing to more responsibility.

In reality, replanting often reduces effort rather than increasing it. New plants are typically easier to manage than struggling ones, and fresh growth often stabilizes watering and nutrient use.

Replacing a tired plant with a new one can simplify care rather than complicate it.

What This Phase Teaches You

Replanting reinforces an important lesson.

Gardens are not static projects with a single correct outcome. They are ongoing systems that adapt, shift, and renew themselves.

Once you experience that renewal firsthand, gardening stops feeling like something that can "end badly." There is always a next step, and it's usually simpler than expected.

Letting the Bed Evolve Naturally

As you replant and extend the season, the bed will begin to look less planned and more lived-in.

That's not a sign of disorder. It's a sign that the system is functioning.

Plants at different stages share space. Soil improves with use. The bed becomes more forgiving, not less.

This evolution is one of the quiet rewards of sticking with it.

What to Do Next

As the season winds down or shifts, attention naturally turns inward rather than outward.

What worked well. What you'd repeat. What you'd skip next time.

In the final chapter, we'll talk about how to carry those observations forward without judgment or pressure, so each season feels easier than the last instead of heavier.

For now, know this.

You don't need to start over to keep going.

You just need to notice when it's time to begin again, in smaller ways.

Chapter 15:

Learning From This Season (Without Beating Yourself Up)

By the time you reach the end of a season, the garden has usually taught you far more than any single instruction ever could, not through dramatic wins or obvious failures, but through repetition, small adjustments, and a gradual sense that things began to work with less effort than they did at the start.

This is often the moment when people feel an urge to evaluate everything at once, replaying decisions, second-guessing timing, and mentally listing what they believe should have gone differently. That instinct is understandable, but it can easily turn unproductive if reflection becomes judgment rather than learning.

This chapter is about slowing that process down and keeping it constructive, so what you take forward is confidence and clarity rather than a sense that you need to do better next time to justify the effort you already made.

Why Reflection So Easily Turns Into Self-Criticism

Gardening outcomes rarely trace back to a single choice, even though it can feel that way when something disappoints.

Weather shifts unexpectedly. Temperatures fluctuate just enough to affect growth. Certain varieties simply fail to thrive in specific microconditions despite reasonable care. None of these factors respond to effort or intention, yet they strongly influence results.

When reflection focuses on identifying mistakes, it tends to assign responsibility where it does not belong. When reflection focuses on patterns, it produces information you can actually use.

Patterns reduce future effort. Blame increases it.

What Is Actually Worth Paying Attention To

You do not need detailed records or a season-long journal to learn effectively from a raised bed.

A few broad observations are usually enough to guide future decisions, especially when they center on effort rather than outcomes alone. Noticing which plants grew steadily with minimal attention, which ones demanded frequent intervention, and which harvests felt rewarding relative to the time invested tells you far more than tracking every success and setback.

Gardens become easier when you repeat what felt easy and quietly stop doing what didn't.

Separating Plant Performance From Your Experience of It

Sometimes a plant grows exactly as expected but feels inconvenient or fussy to manage. Other times, a plant struggles or produces modestly, yet you enjoy having it in the bed and interacting with it.

Both responses matter.

A garden that produces food but consistently feels like work is unlikely to last, regardless of how productive it looks on paper. A garden that balances usefulness with enjoyment is far more sustainable, even if yields are not maximized.

You are allowed to prioritize ease, satisfaction, and rhythm over theoretical efficiency, and doing so often leads to better results anyway.

Why Comparison Interferes With Learning

It is difficult to reflect clearly when comparison enters the picture.

Looking at other gardens, whether online or nearby, invites assumptions about what should have happened in your own space. What those comparisons rarely account for are differences in climate, light, time availability, priorities, and even personality.

Learning works best when it stays anchored in your conditions rather than someone else's outcome. Your garden does not need to resemble anyone else's to be successful.

It only needs to function well for you.

Small Adjustments Carry More Weight Than Big Resets

After a first season, there is often a temptation to overhaul everything, replacing plants, changing layouts, adding beds, or redesigning systems from the ground up.

Sometimes expansion is exciting and appropriate, but more often, small adjustments deliver better results with far less effort. Choosing a different variety, spacing plants slightly differently, improving mulch coverage, or making watering access easier can quietly improve an entire season without increasing complexity.

Raised beds reward continuity. There is no advantage in starting from scratch when incremental refinement works just as well.

Letting Go of What Didn't Earn Its Place

Some plants will not justify their space, and recognizing that is part of developing confidence rather than admitting defeat.

A plant that grew slowly, attracted persistent pests, or produced less than expected is not a failure. It is information. Removing it from future plans simplifies decisions and reduces maintenance without reducing satisfaction.

Gardens improve when you stop trying to make everything work and start choosing plants that are well suited to your space and your preferences.

Why Familiarity Matters More Than Mastery

At this stage, you may notice that tasks which once felt intimidating now feel straightforward, not because you learned every detail, but because repetition removed uncertainty.

Planting feels simpler. Watering decisions feel clearer. Harvesting feels less tentative. That ease comes from familiarity with the process rather than complete understanding of it.

Gardening confidence grows through exposure, not expertise, and each season quietly reinforces that familiarity.

Preparing for the Next Season Without Pressure

There is no need for elaborate planning or early optimization.

A short mental list is usually enough to guide the next round: what you would grow again without hesitation, what you would skip, and one small adjustment you would try differently.

Keeping planning light prevents it from becoming another obligation and preserves the sense that gardening fits into your life rather than competing with it.

Reframing a Common End-of-Season Thought

Many people reach the end of a season thinking they could have done better.

A more useful reframing is noticing how much easier decisions felt by the end than they did at the beginning, and how much less effort was required to achieve similar or better results.

Ease is the metric that compounds over time.

Allowing the Garden to Pause

There is no requirement to garden continuously.

If the bed rests for a period, nothing meaningful is lost. Soil structure remains intact. Organic matter continues to break down. Experience remains available whenever you choose to return.

Gardening can pause without being abandoned, and that flexibility is part of what makes raised beds so forgiving.

What This Season Really Gave You

Even a modest garden offers more than produce.

It builds patience.

It reinforces trust in simple systems.

It shows that effort does not need to be constant to be effective.

Those lessons extend beyond the bed itself and make future seasons feel lighter rather than heavier.

Where This Leaves You

If you continue gardening, the next season will feel easier, not because conditions will suddenly be perfect, but because you will recognize what matters and what can be ignored.

If you stop for a while, the knowledge remains intact and ready when you return.

Either outcome is valid, and neither erases what you learned.

Closing Thought

Gardening success is not measured by flawless beds or uninterrupted harvests, but by how willing you are to return to the process without dread or self-blame.

If this season felt manageable, even imperfectly so, then it worked.

And that is enough to build on.

Bonus:

The Low-Effort Gardener's Toolkit

A practical reference you can return to without thinking

This section exists so you don't have to rely on memory, confidence, or motivation to keep your raised bed going. When something feels uncertain, busy, or slightly off, these tools let you act without overthinking.

You can treat this section as a reference manual, not a reading assignment.

Tool 1: The 30-Day Quick Start Calendar (Expanded)

This is not a productivity plan. It's a **reality map**, so you don't panic when the garden behaves normally.

The 30-Day Quick Start Calendar

A reality map for your first month, not a to-do list

WEEK 1. SETTLE & OBSERVE

Days 1-7

MON	TUE	WED	THU	FRI	SAT	SUN
Soil settles	All quiet	Soil settles	All quiet	Soil settles	All quiet	Soil settles

What you'll notice:

- Soil settles and compacts slightly
- Seeds are invisible
- Seedlings look unchanged

Water only when soil below the surface feels dry

What this week is for: Letting roots and soil stabilize.

No feeding, no adjusting, no fixing. **DO NOT** panic about "nothing happening"

WEEK 2. UNEVEN BEGINNINGS

Days 8-14

MON	TUE	WED	THU	FRI	SAT	SUN
Lagging	Sprout	Lagging	Sprout	Lagging	Sprout	Sprout

What you'll notice:

- Some seeds emerge, others lag
- Seedlings may droop briefly, especially in heat
- Growth looks uneven and messy

Keep watering steady

What this week is for: patience and restraint.

DO NOT dig, replant"just to check", fertilize

WEEK 3. MOMENTUM APPEARS

Days 15-21

MON	TUE	WED	THU	FRI	SAT	SUN

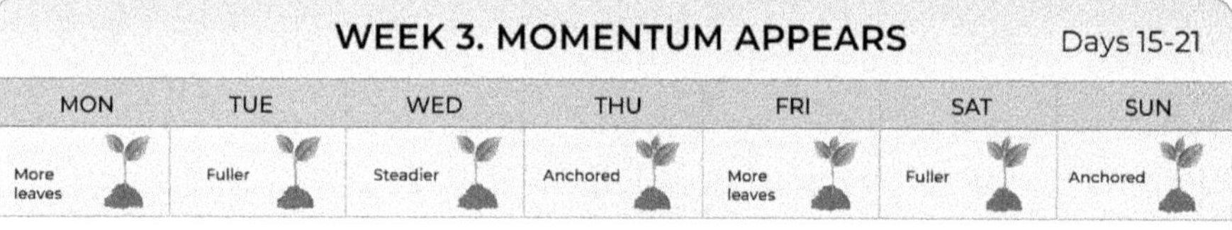

What you'll notice:

- New leaves appear more regularly
- Plants look anchored, not flimsy
- Growth is easier to notice

Observe. Light harvesting possible if appropriate.

What this week is for: trusting the garden.

DO NOT overfertilize "just because it's working"

WEEK 4. STABILITY

Days 22-30

- Bed looks fuller and more intentional
- Watering decisions feel clearer
- Plants tolerate small mistakes better
- Confidence increases naturally

What this week is for: settling into maintenance.

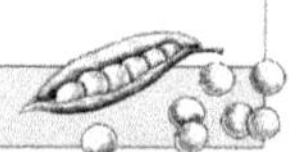

If your garden followed this timeline, nothing went wrong.

Most beginner mistakes come from **reacting too early**, not from neglect.

Week 1: Settle and Observe

- Soil settles and compacts slightly
- Seeds are invisible, seedlings look unchanged
- Water only when soil below the surface feels dry
- No feeding, no adjusting, no fixing

What this week is for: letting roots and soil stabilize

What not to do: react to the lack of visible progress

Week 2: Uneven Beginnings

- Some seeds emerge, others lag behind
- Seedlings may droop briefly, especially in heat
- Growth is inconsistent across the bed
- Continue steady watering

What this week is for: patience and restraint

What not to do: dig, replant prematurely, or add fertilizer

Week 3: Momentum Appears

- New leaves form more regularly
- Plants look more anchored

- Growth becomes easier to notice
- Light harvesting of fast greens may be possible

What this week is for: observation without interference

What not to do: assume faster growth means more input is needed

---

Week 4: Stability

- Bed looks fuller and more intentional
- Watering decisions feel clearer
- Plants tolerate small mistakes better
- Confidence increases naturally

What this week is for: settling into maintenance rather than setup

---

Tool 2: The Weekend Setup Checklist

The Weekend Setup Checklist

To prevent unnecessary scope creep.

ABSOLUTELY REQUIRED

- [ ] Raised bed in a reasonably sunny, convenient spot
- [ ] Soil filled and watered
- [ ] Seeds or seedlings planted
- [ ] Access to water

If this is done, you are gardening.

Helpful if you have time

- [ ] Mulch added to soil surface
- [ ] Hose, watering can, or simple irrigation ready
- [ ] Basic snips or scissors

These reduce effort later, not success now.

Nice but not necessary

- Perfect alignment
- Decorative borders
- Upgraded tools
- Specialty products

If something delays planting, skip it.

Tool 3: Shopping Lists by Budget

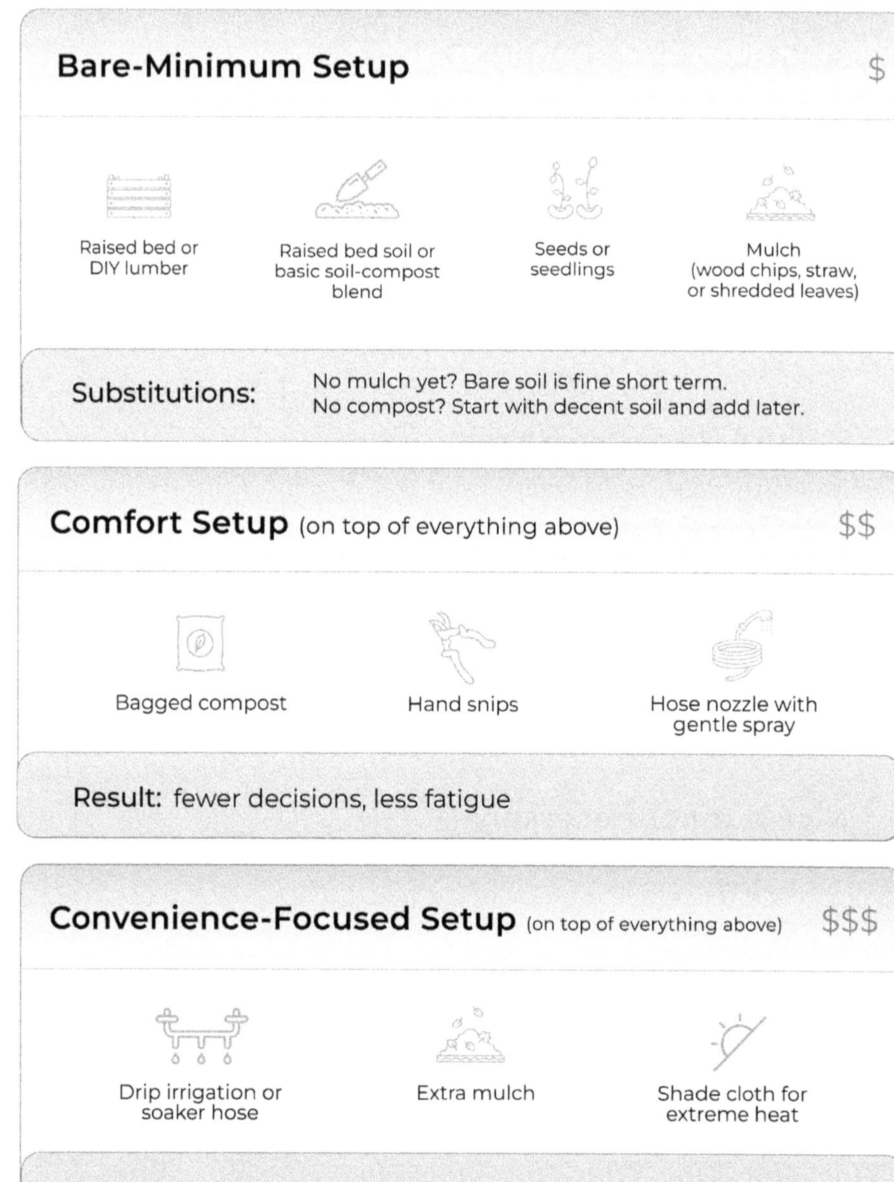

Tool 4: One-Page Planting Cheat Sheet

Planting Cheat Sheet

SEEDS

Plant shallower than expected

Keep surface evenly moist until sprouting

Expect uneven germination

SEEDLINGS

Plant at original soil depth

Water immediately and thoroughly

Expect a pause in growth

SPACING

Raised beds tolerate closer spacing

Full soil coverage reduces weeds

Airflow matters later, not immediately

If choosing between exact spacing and simplicity, ***choose simplicity***.

Tool 5: One-Page Watering Guide (Expanded)

One-Page Watering Guide

Decision Rule

Check **soil depth**, not surface appearance.

Moist
below the surface

▼

Dry
below the surface

▼

Watering Style

- **Fewer, deeper** waterings beat frequent light watering
- **Let soil dry** slightly between watering
- Avoid constant dampness

Seasonal Adjustments

- **Hot weather:** increase frequency
- **Cool/cloudy weather:** reduce frequency
- **Mulched beds:** water less often

Watering consistency matters more than precision.

Tool 6: One-Page Troubleshooting Chart (Expanded)

One-Page Troubleshooting Chart

If Plants Look Off

1. Has watering been consistent?
2. Were they planted recently?
3. Is more than one plant affected?

If Leaves Are Damaged

- **Growth continuing:** ignore
- Damage increasing rapidly: **inspect**

If Growth Is Slow

- Early stage: **wait**
- Post-harvest slowdown: **consider feeding**
- Whole bed affected: **check watering first**

If Unsure...

Waiting is usually safer than acting.

Tool 7: The Feeding Decision Guide and Tool 8: The Replanting Decision Guide

The Feeding Decision Guide

An extra layer to prevent overfeeding

Healthy growth

Do nothing

Slower regrowth after harvest

Light feeding may help

Stress or discoloration

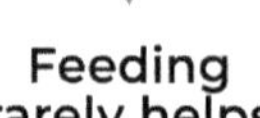

Feeding rarely helps

The Replanting Decision Guide

So you don't hesitate or overcommit

Replace a Plant When

- It has stopped producing
- It looks stressed despite care
- It blocks light or space

Keep a Plant When

- It is still producing at any level
- It looks healthy overall

Mixed stages are normal and productive.

Tool 9: The "Ignore This" List and Tool 10: The Confidence Reset Tool

The "Ignore This" List

Permission matters

- Perfect spacing
- Online comparison photos
- One-size-fits-all calendars
- Advice that adds stress without reducing effort

If advice doesn't make gardening easier, it's optional.

The Confidence Reset Tool

For the days you doubt yourself

Ask yourself:

- Is anything clearly dying?
- Has growth completely stopped?
- Has something changed dramatically?

If the answer is **NO**, you're doing fine.

Gardens do not collapse quietly. They signal clearly when help is needed.

Final Note to the Reader

These tools exist so you don't have to become "a gardener" to grow food.

You don't need ambition.

You don't need expertise.

You don't need constant attention.

You need a system that works even when you're busy, distracted, or unsure.

That's what this toolkit is for.

Conclusion.

You Were Never Bad at Gardening

If there is one thing worth carrying forward from everything you've read here, it's this: raised bed gardening was never failing you, and you were never failing it, you were simply being asked to work with a system that was louder, more complicated, and more demanding than it needed to be, and once that noise is stripped away, what remains is something far calmer, far more forgiving, and far more compatible with a real life that already asks enough of you.

Most people don't give up on gardening because they don't care, they give up because the process makes them feel behind before they've even started, as though success belongs to people with more time, better instincts, or a natural ability they somehow missed, when in reality the difference is almost always about structure and expectations rather than effort or intelligence.

Raised beds change that equation quietly, not by making gardening perfect, but by making it manageable, by reducing the number of things that can go wrong at once, by containing chaos just enough that learning can happen without constant correction, and by giving you a place where plants can succeed even when your attention isn't flawless.

What you've done by following this approach, whether you realized it or not, is replace urgency with rhythm, reaction with observation, and self-doubt with a set of decisions that don't need revisiting every day, and that shift alone is what allows confidence to take root, not because everything goes right, but because nothing collapses the moment it goes slightly wrong.

You may still lose a plant from time to time, you may still misjudge watering during a heat wave, you may still try something that doesn't earn its place in the bed, and none of that disqualifies you from success, because gardening was never meant to reward perfection, it rewards consistency, patience, and the willingness to let things be imperfect long enough to grow.

If your bed produces food, even modestly, if it fits into your life without becoming another obligation, if it feels easier to return to than to avoid, then it is doing exactly what it's supposed to do, and you are doing exactly what you need to do to keep it working.

And if at some point you step away for a while, whether because of weather, work, health, or simple lack of energy, nothing meaningful is lost, the soil remains improved, the understanding remains available, and the system will be ready whenever you decide to come back to it, without judgment or penalty.

Before you close this book, take a moment to notice how different gardening feels now compared to when you started, not because you memorized techniques or mastered terminology, but because the process no longer feels fragile, and that is the real outcome worth protecting. For many people, that shift shows up as less overthinking, fewer reactive fixes, and more trust in simple routines.

If this book helped you feel calmer, clearer, or more capable of growing food in raised beds without turning it into a project you had to manage constantly, I'd like to ask one small favor, not for me, but for the next person standing where you were, uncertain and quietly wondering if it's worth trying again.

No Overthinking Rule #7

Progress beats optimization. Always.

Many readers find that even a short note about what changed for them makes a real difference to someone else reading reviews while deciding whether to start.

Leaving a short, honest review, even just a few sentences about what helped or what felt different this time, helps other beginners recognize themselves in your experience and gives them permission to start without feeling foolish for not knowing everything yet, and in that way, something as simple as your words can remove a barrier that stops many people before they ever plant a single seed. It does not need to be polished or detailed to be useful.

Just scan this QR code to access the review link:

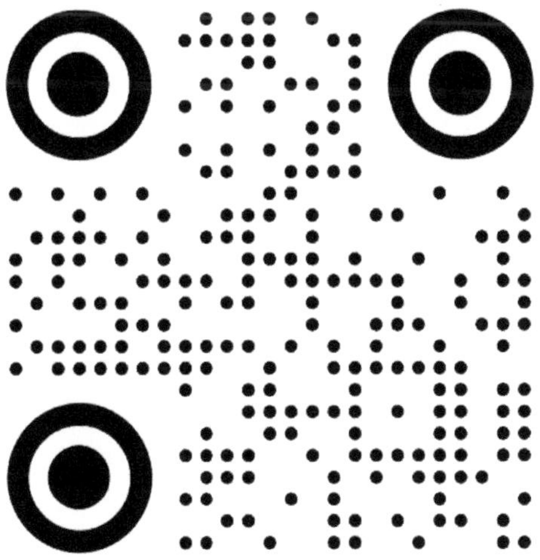

Gardening doesn't need more experts, it needs more people who are willing to say, honestly, that good enough really does grow food, and that starting small, staying simple, and allowing yourself to learn slowly is not a compromise, it's the reason this works at all.

If this approach helped you succeed where things felt complicated before, that story is worth sharing.

Selected Sources & Further Reading

This book is grounded in established horticultural research and extension guidance, adapted for simplicity, realism, and everyday use.

Bachman, G. R., & Metzger, J. D. (2017). *The gardener's guide to raised bed gardening.* Cool Springs Press.

Brennan, E. B., & Boyd, N. S. (2012). *Best management practices for raised bed vegetable production.* University of California Agriculture and Natural Resources.

Cornell University Cooperative Extension. (2020). *Raised bed gardening.* https://gardening.cals.cornell.edu

Clemson Cooperative Extension. (2019). *Home vegetable gardening.* https://hgic.clemson.edu

Davis, J. G., & Whiting, D. (2013). *Gardening in raised beds.* Colorado State University Extension. https://extension.colostate.edu

Flint, M. L. (2012). *Integrated pest management for home gardeners.* University of California Agriculture and Natural Resources.

Royal Horticultural Society. (2021). *RHS gardening through the year.* Dorling Kindersley.

University of Minnesota Extension. (2022). *Raised beds.* https://extension.umn.edu

Washington State University Extension. (2020). *Home vegetable gardening in raised beds.* https://extension.wsu.edu

www.ingramcontent.com/pod-product-compliance
Ingram Content Group UK Ltd.
Pitfield, Milton Keynes, MK11 3LW, UK
UKHW021036270726
13967UKWH00013B/2816

9 781764 459419